SUSTAINABLE LANDMARKS

可持续性地标建筑（下）

石大伟 主编

中国林业出版社

TANGO DISCO

Beijing, China, 2006-2008

地点　北京，中国
项目　娱乐场所
客户　北京糖果娱乐有限公司
规划　2006年
建设　2007年-2008年
成本　180,000欧元
建筑面积　2,000平方米
建筑正面面积　600平方米
体量　8,000立方米
承包商　Beijing Hong HengJi Wall Engineering and Decoration Ltd.

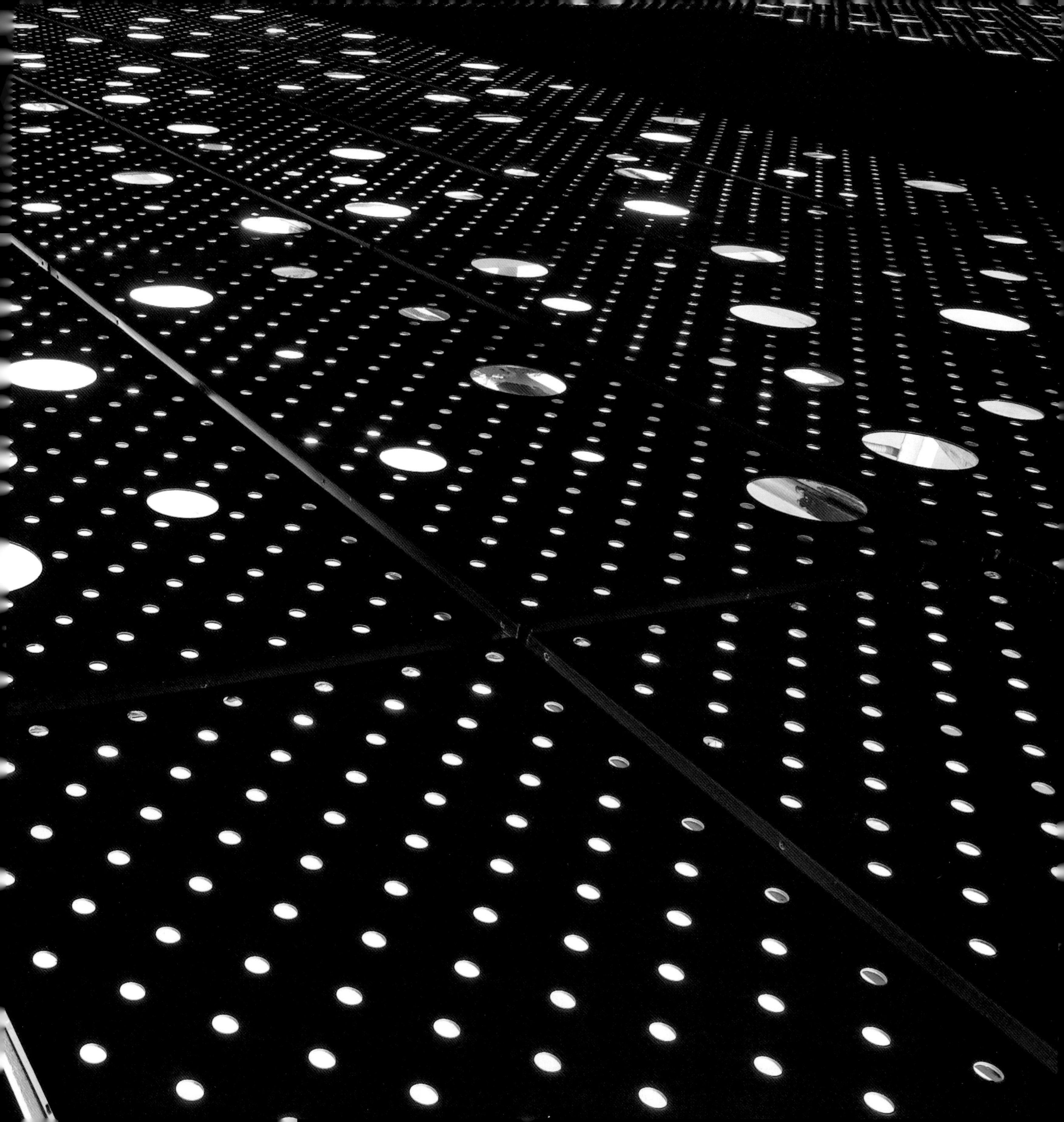

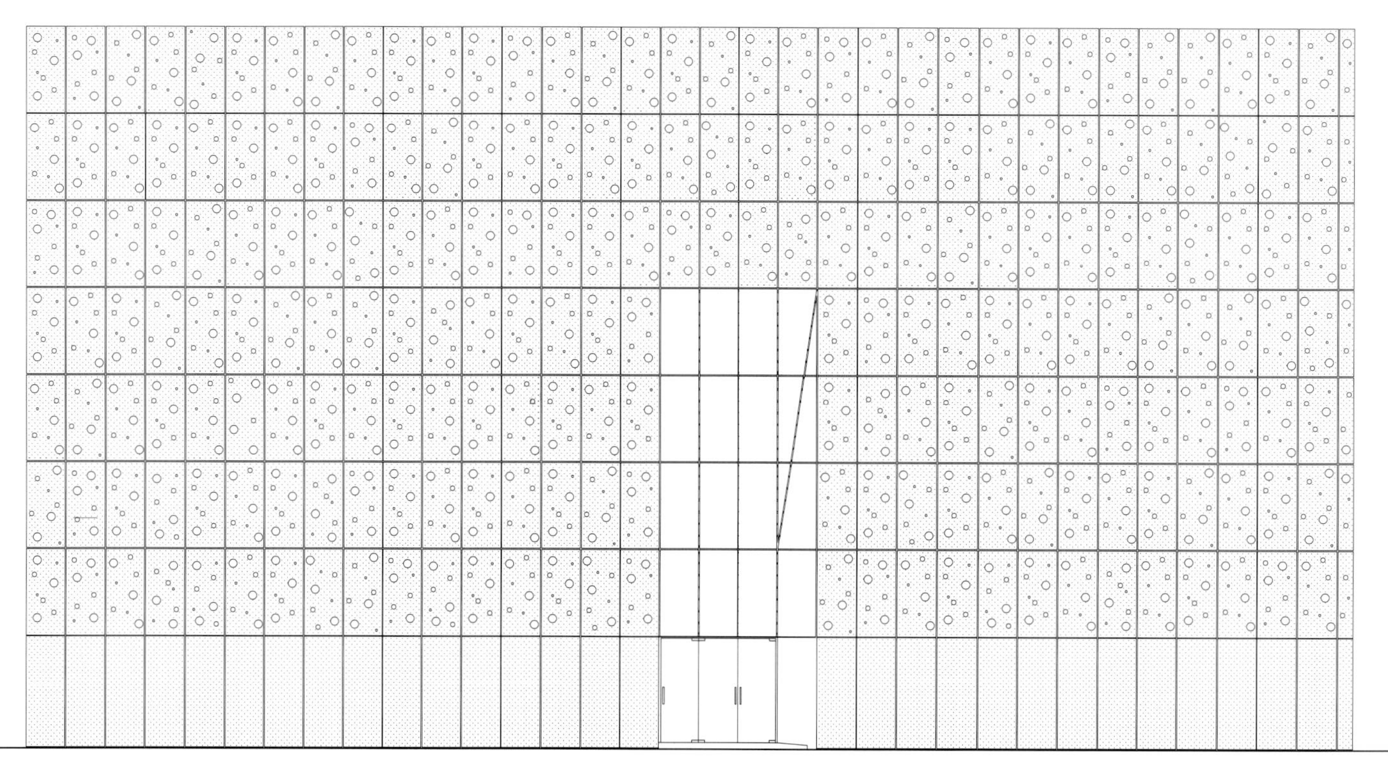

main elevation

主正视图

0 1 2 m

北京市民族文化交流中心
北京人才市场

24小时营业
第一家
金鼎轩

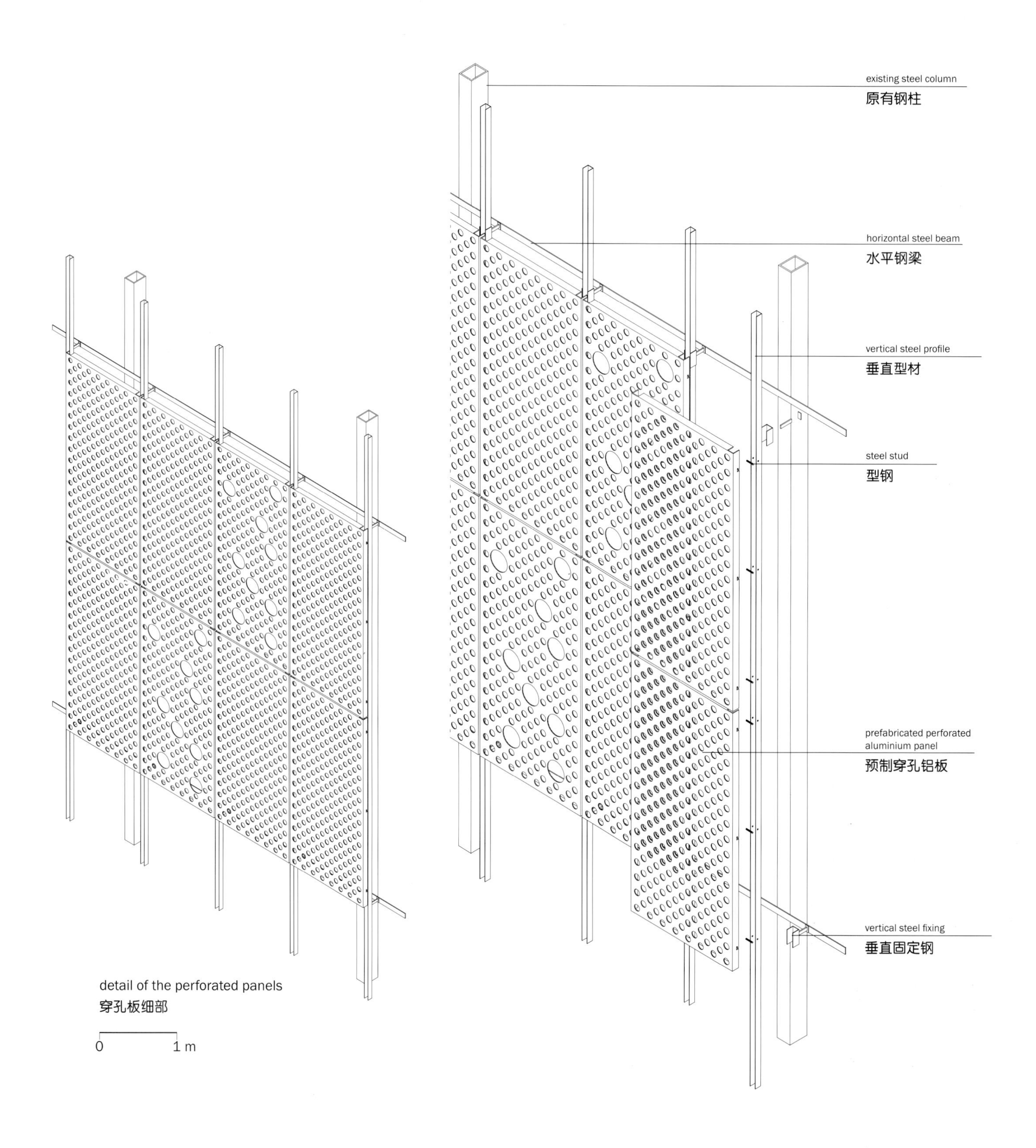

detail of the perforated panels
穿孔板细部

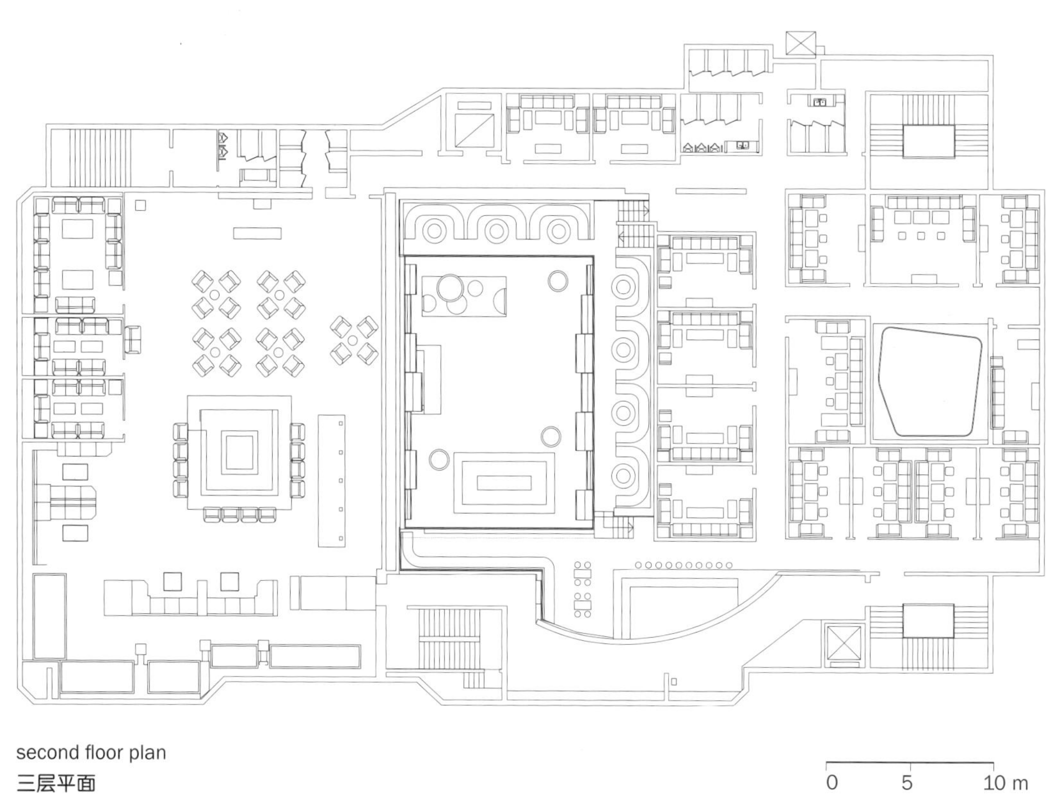

second floor plan

三层平面

MUNICIPALITY OFFICES AND CULTURAL CENTER

Figline Valdarno, Florence, Italy
2006 - under construction

地点　Figline Valdarno，佛罗伦萨，意大利
项目　市政府办公楼和文化中心
客户　Comune di Figline Valdarno
结构　Studio G.T.A.
系统　M&E S.r.l.
规划　2006年
建设　在建
成本　4,000,000欧元
建筑面积　3,010平方米
承包商　CFC costruzioni S.r.l.

site plan
位置图

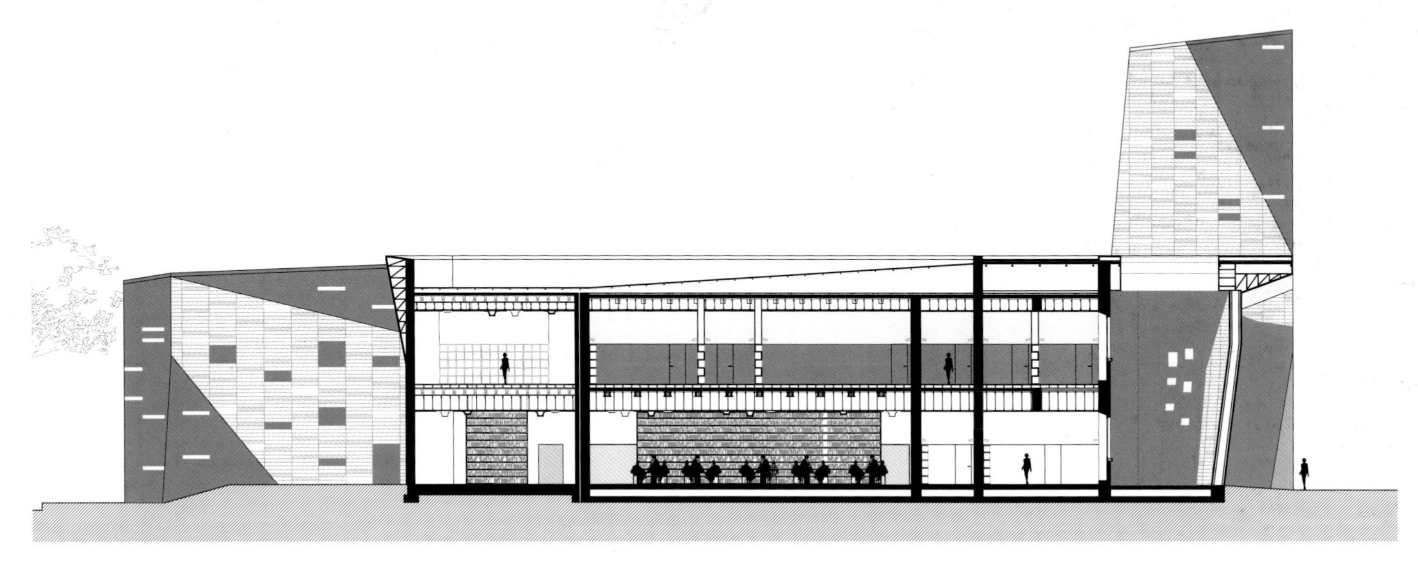

longitudinal section

纵剖面

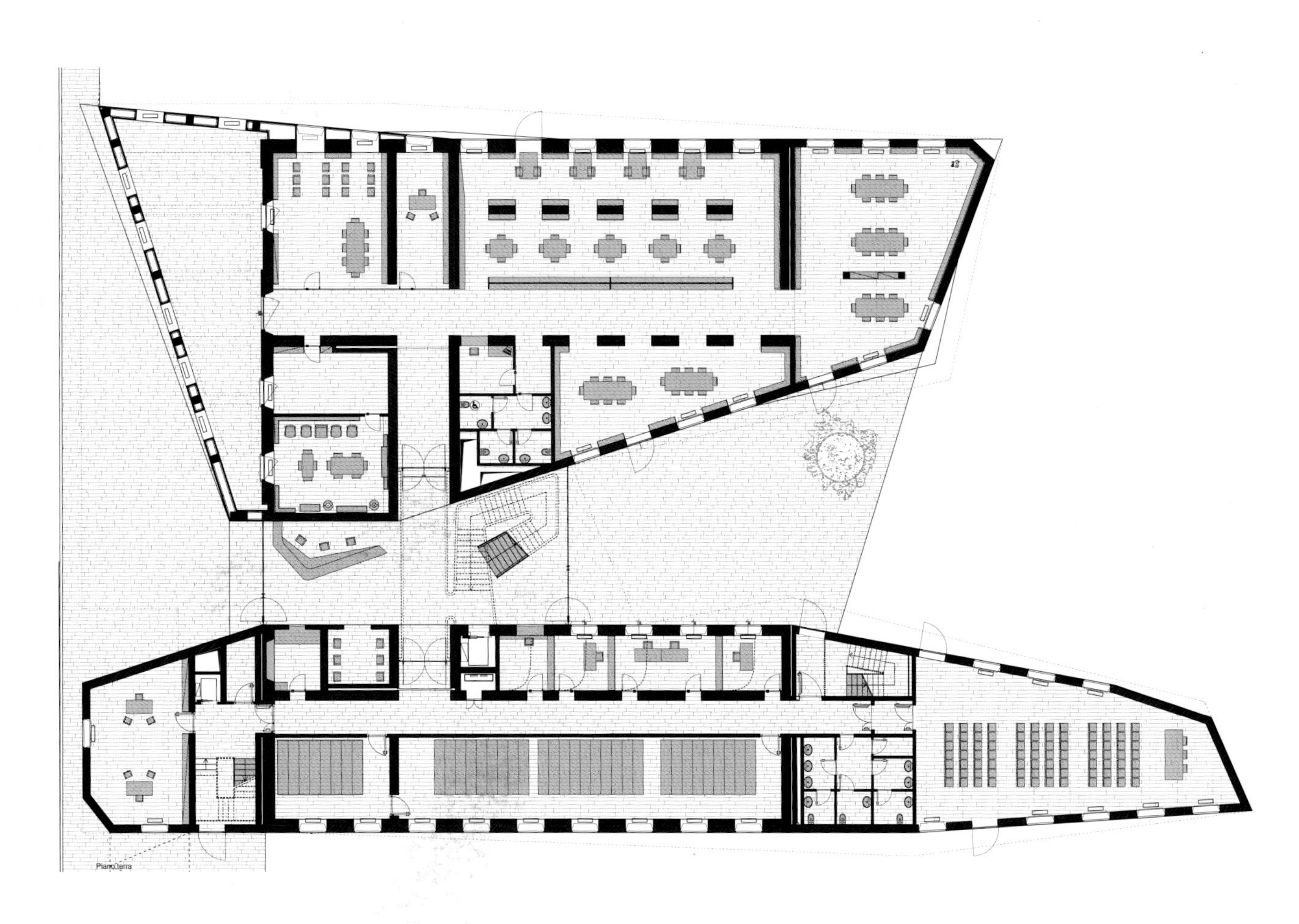

ground floor plan

一层平面

0 5 10 m

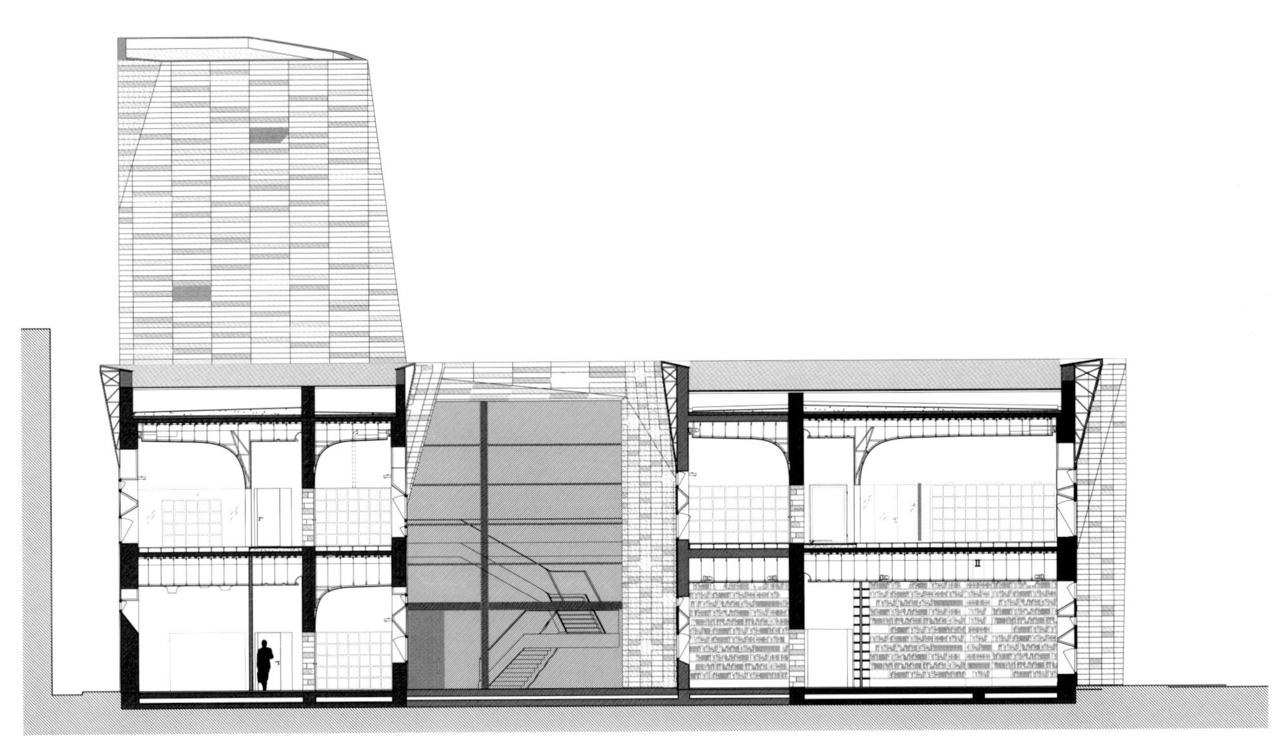

longitudinal section
纵剖面

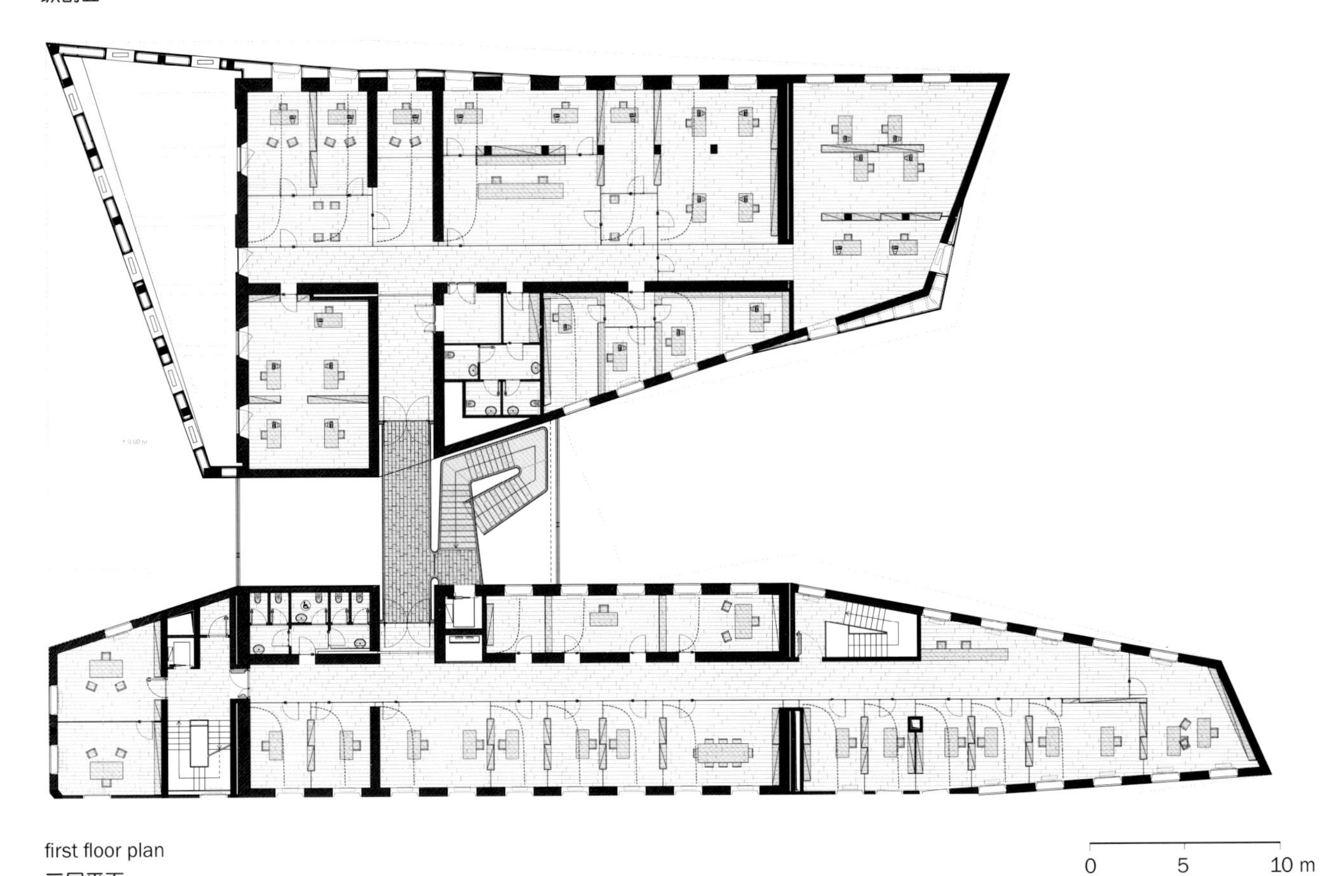

first floor plan
二层平面

FIRENZE NOVA RESIDENTIAL COMPLEX

Florence, Italy, 2006-2010

地点 佛罗伦萨，意大利
项目 住宅建筑群
客户 S.I.IMM. Firenze S.r.l.
结构 Vega Ingegneria - Lorenzo Checcucci
系统 Studio Carbone
规划 2006年-2007年
建设 2007年-2010年
成本 13,234,000欧元
占地面积 3,300平方米
建筑面积 14,340平方米
停车面积 1,805平方米
体量 40,460立方米
承包商 Q5 S.r.l.

TEREX
Genie S-105

cross section
横剖面

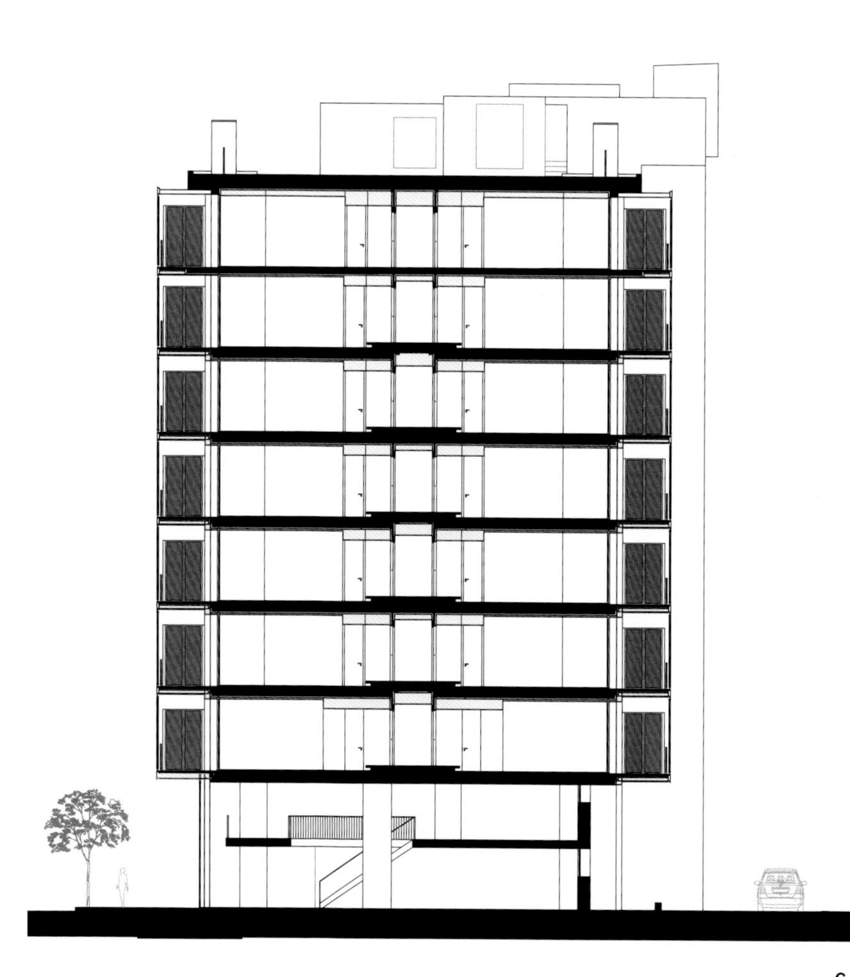

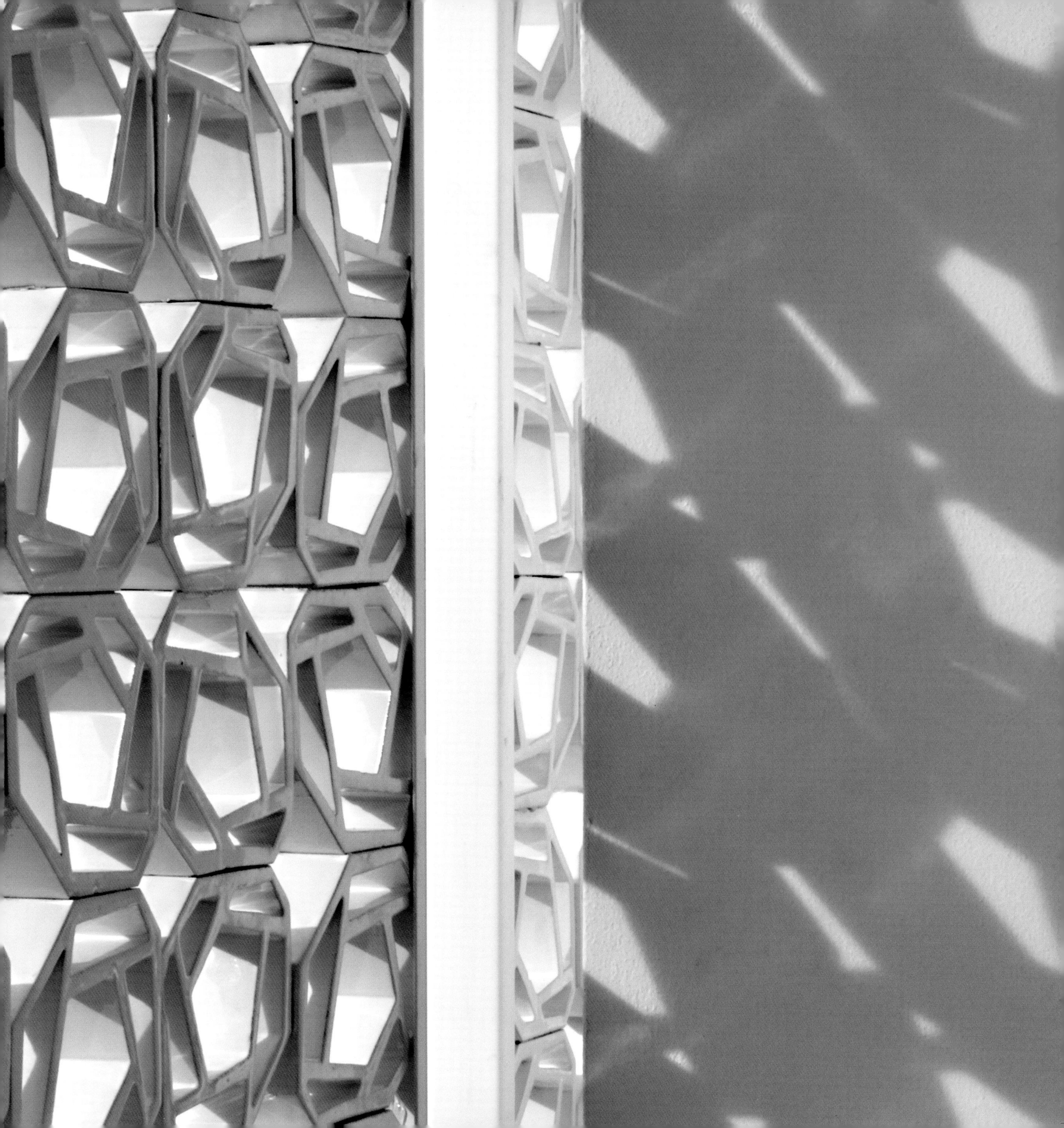

main elevation
主正视图

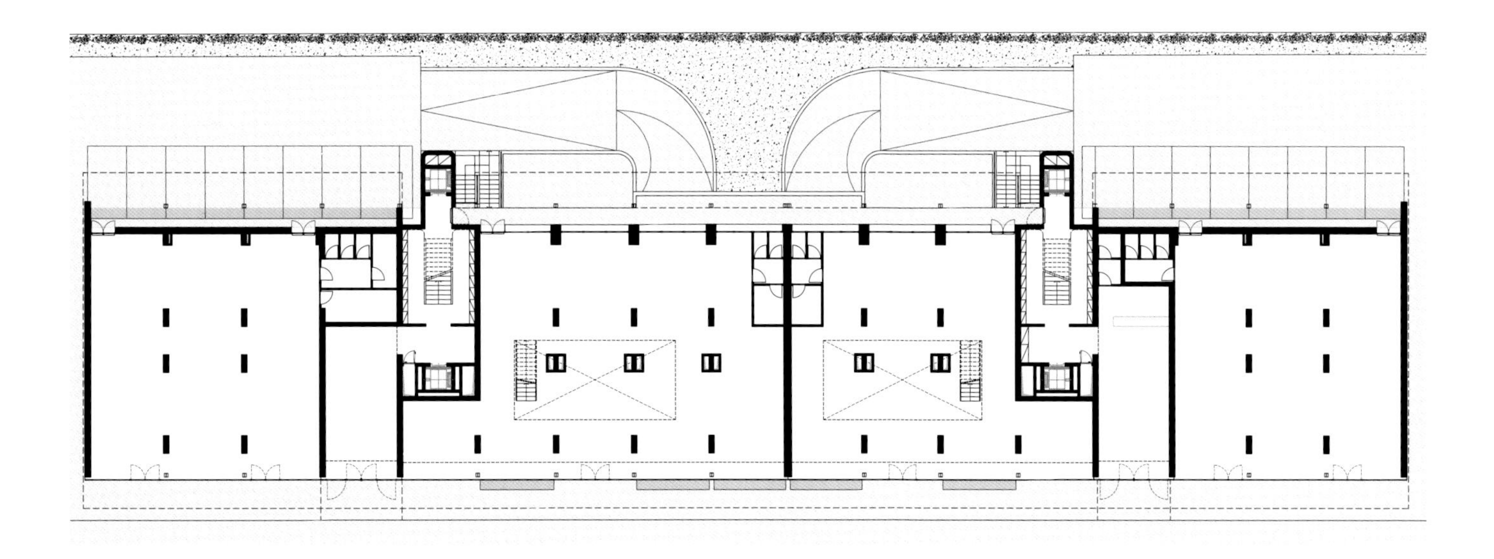

ground floor plan
一层平面

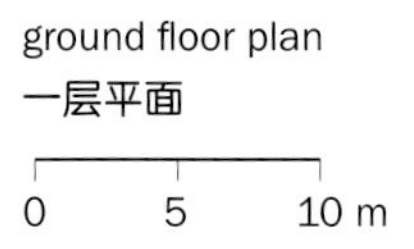

prototype of the element made of glazed clay
粘土釉面部件原型

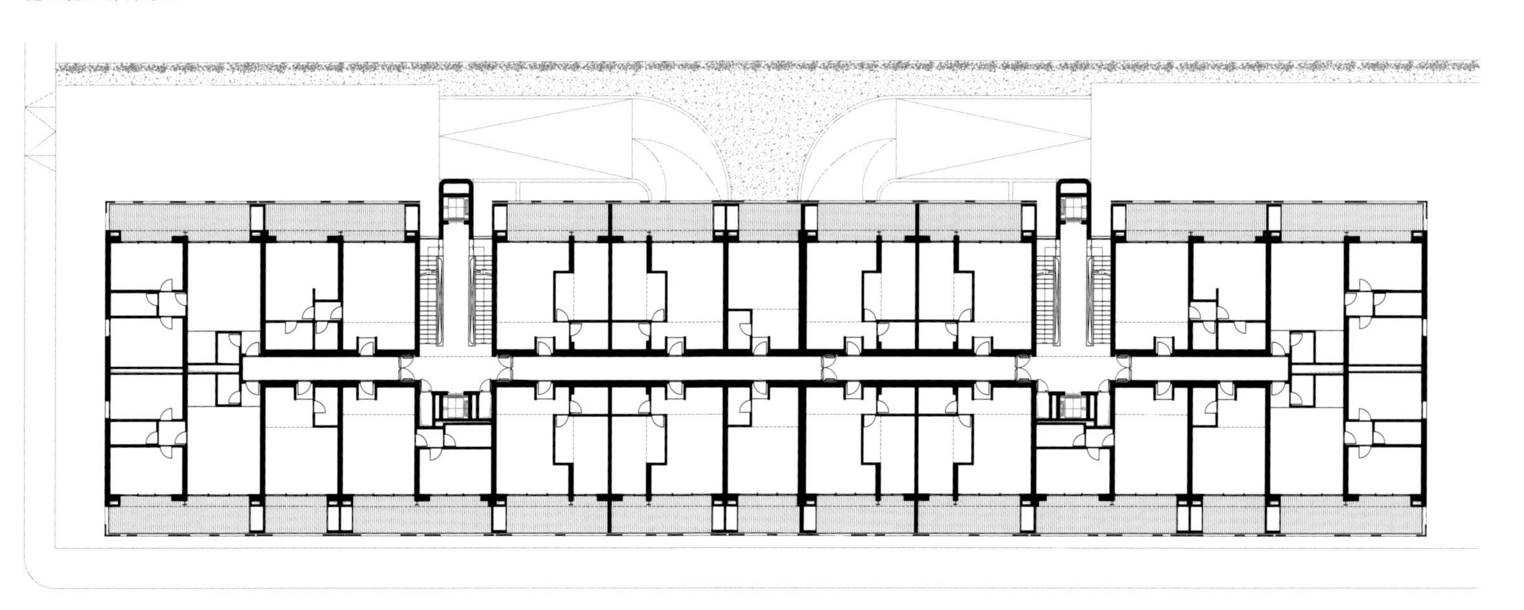

type plan
平面类型

0 5 10 m

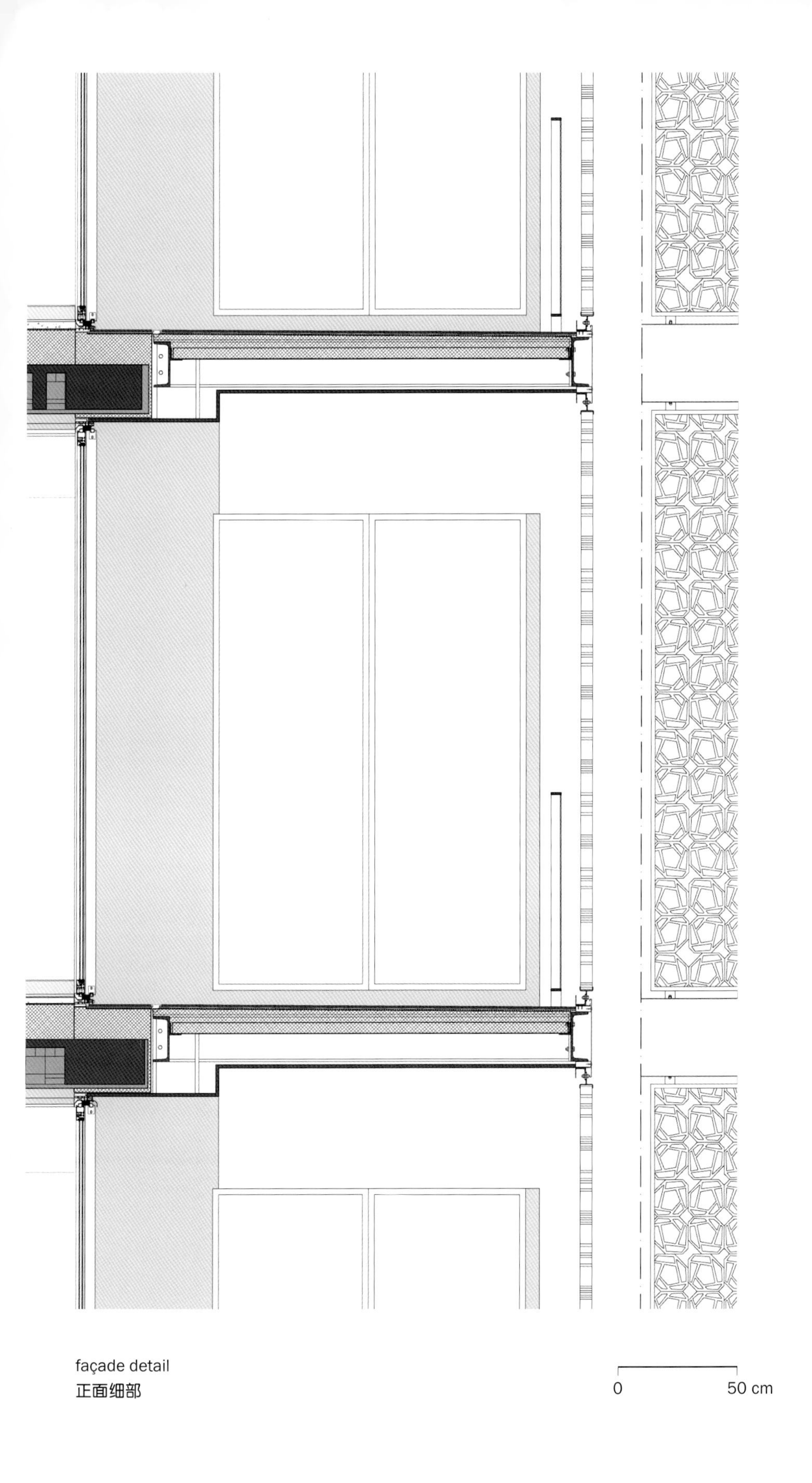

façade detail
正面细部

UBPA B3-2 PAVILION AT SHANGHAI WORLD EXPO 2010

Shanghai, China, 2007-2010

地点　上海，中国
项目　展馆
客户　World Expo Shanghai 2010 Holding Company
结构　Favero&Milan Ingegneria
系统　Favero&Milan Ingegneria
规划　2007年
建设　2007年-2010年
成本　2,000,000欧元
占地面积　3,000平方米
建筑面积　2,000平方米
承包商　Shanghai Construction Company

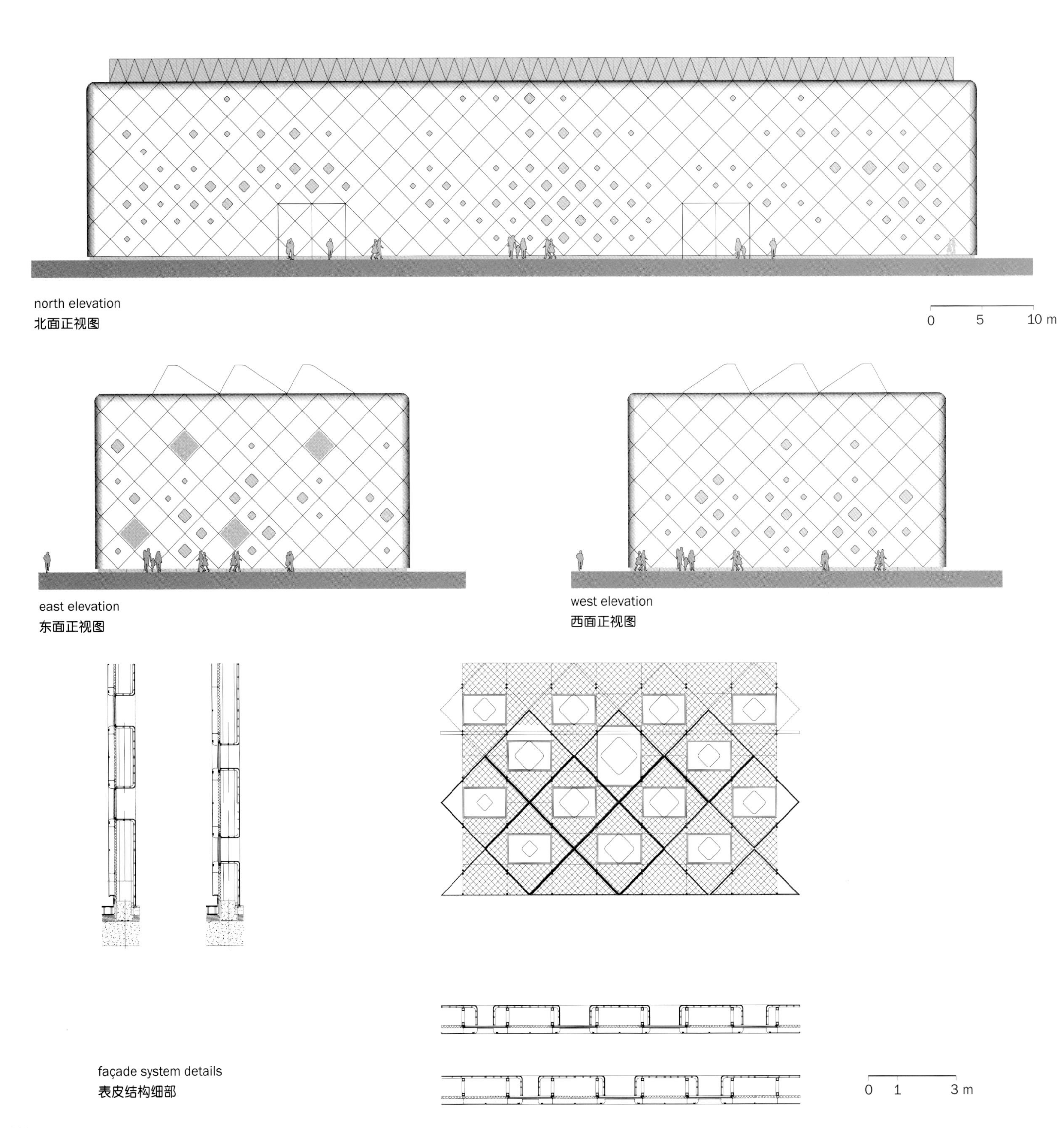

north elevation
北面正视图

east elevation
东面正视图

west elevation
西面正视图

façade system details
表皮结构细部

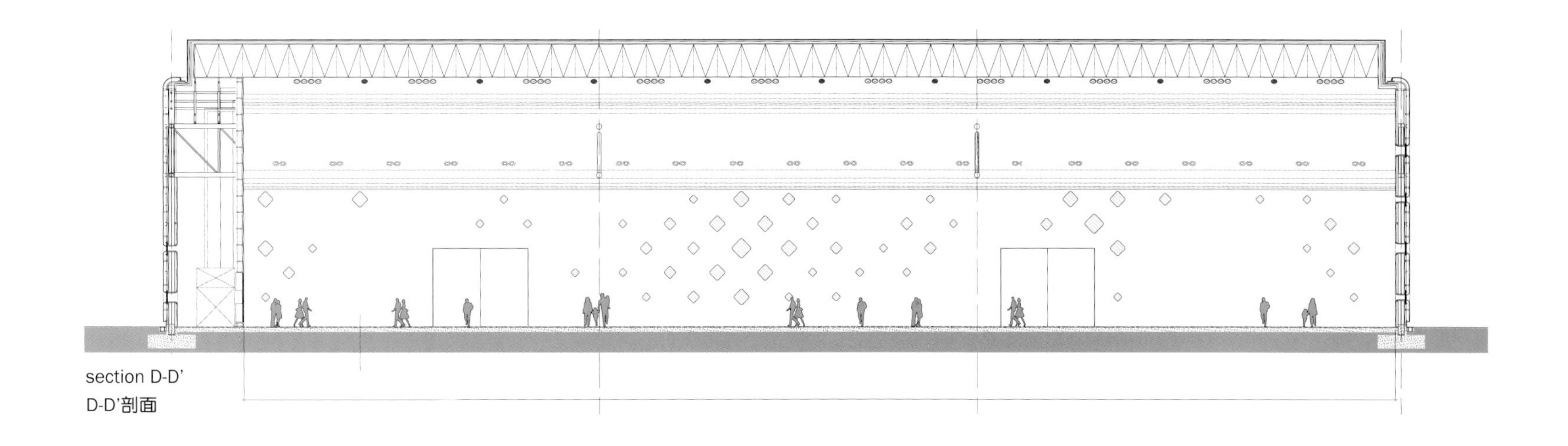

section D-D'
D-D'剖面

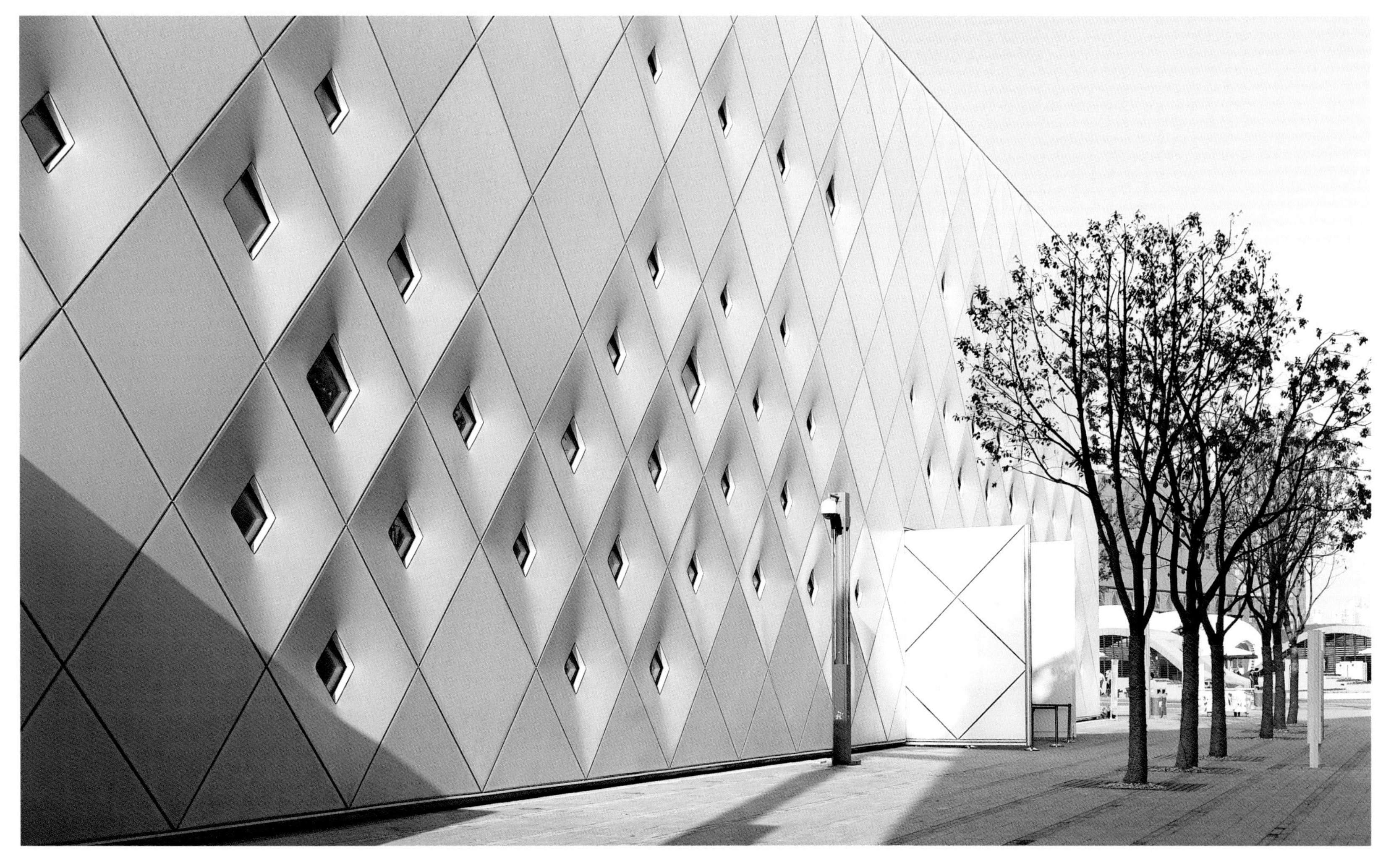

E8
城市最佳实践区
城市广场

案例联合馆3-2
Case joint pavilion 3-2

案例联合馆3-2
Case joint pavilion 3-2

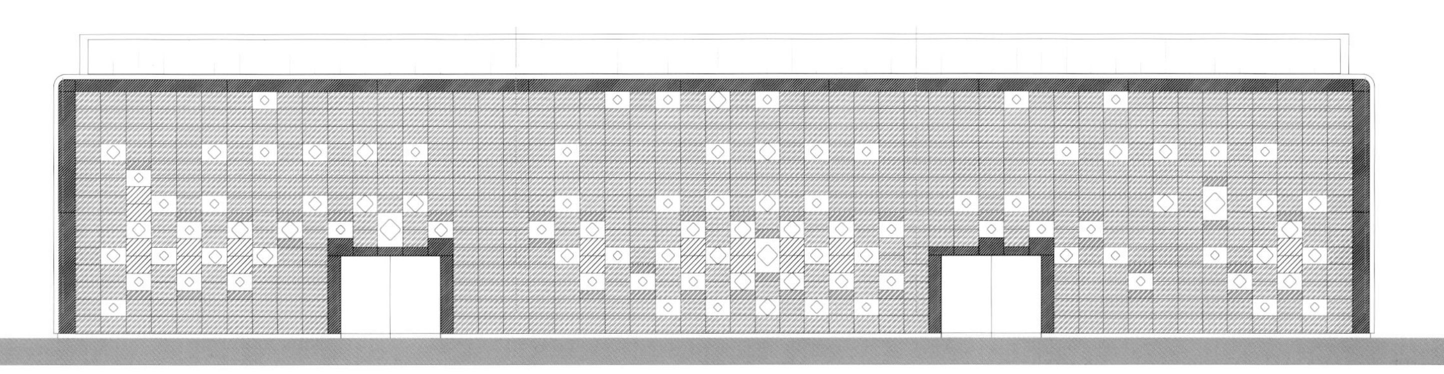

inner north elevation
北面内部正视图

0 5 10 m

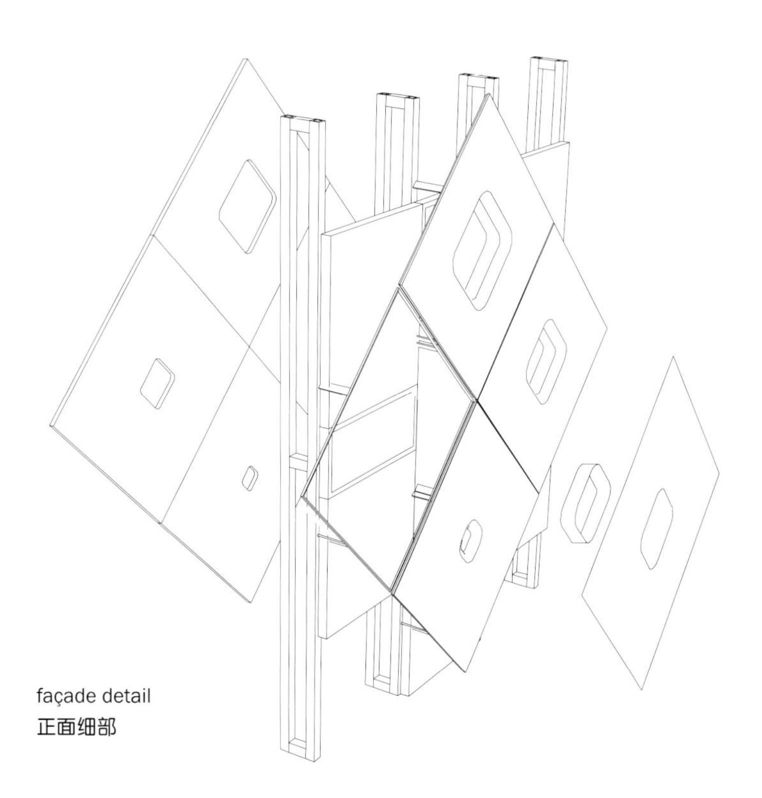

façade detail
正面细部

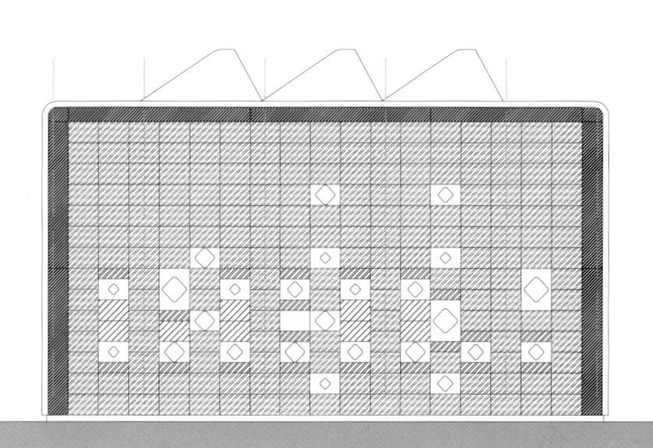

inner west elevation
西面内部正视图

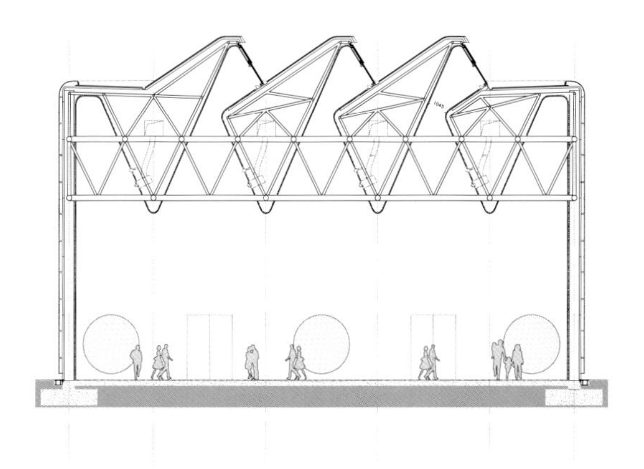

cross section
横剖面

EX MANIFATTURA TABACCHI

multipurpose complex
Cava dei Tirreni, Salerno, Italy
2007 - in progress

地点 Cava dei Tirreni，萨勒莫，意大利
项目 住宅和酒店建筑群
客户 Manifatture Sigaro Toscano S.r.l.
结构 Favero&Milan Ingegneria
系统 StudioTi
规划 2007年
建设 在建
占地面积 13,482平方米
建筑面积 13,310平方米

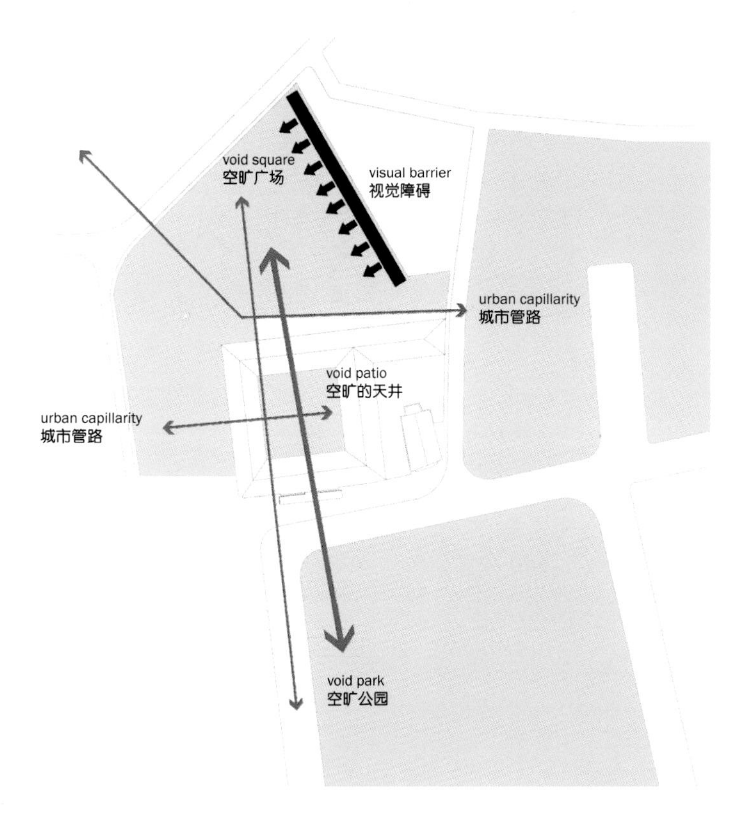

pre-existence
原有的建筑

phase I: sheds demolition
第一阶段：大棚拆除

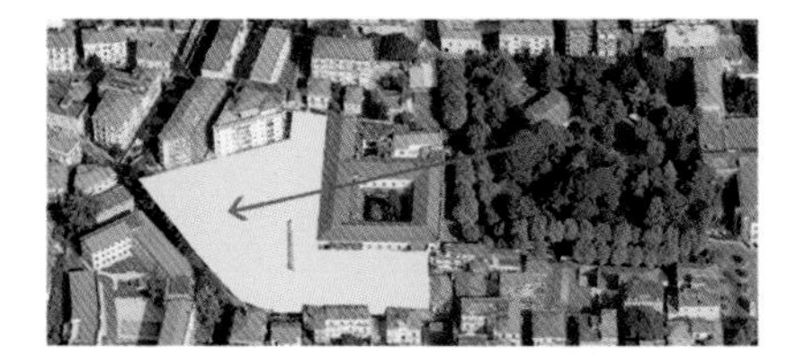
phase II: permeability
第二阶段：渗透

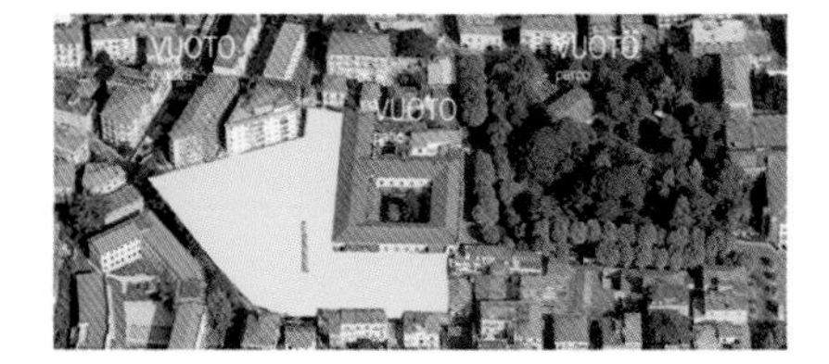

phase III: sequence of voids
第三阶段：空间秩序

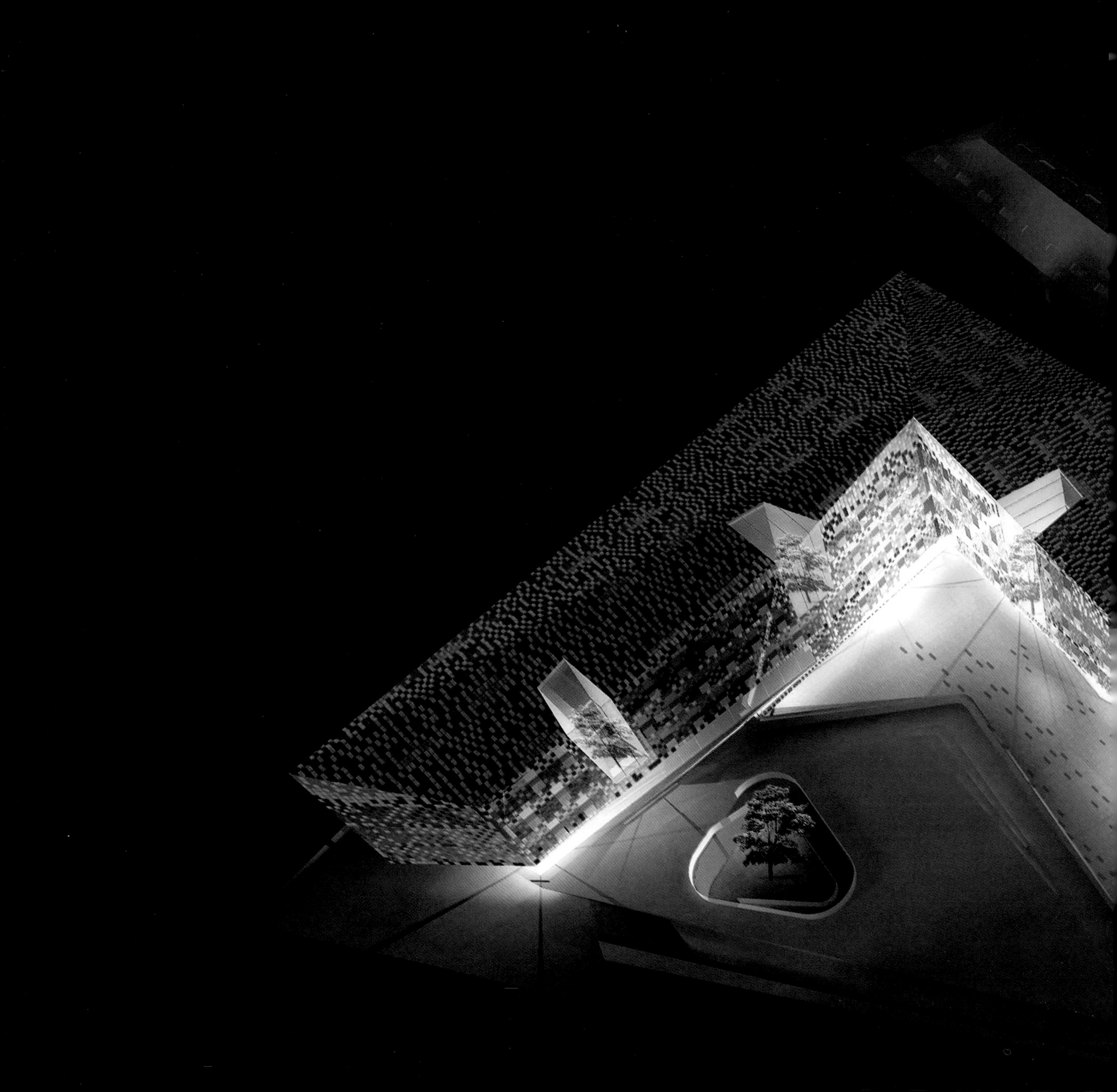

plan level +4.95

高度+4.95平面

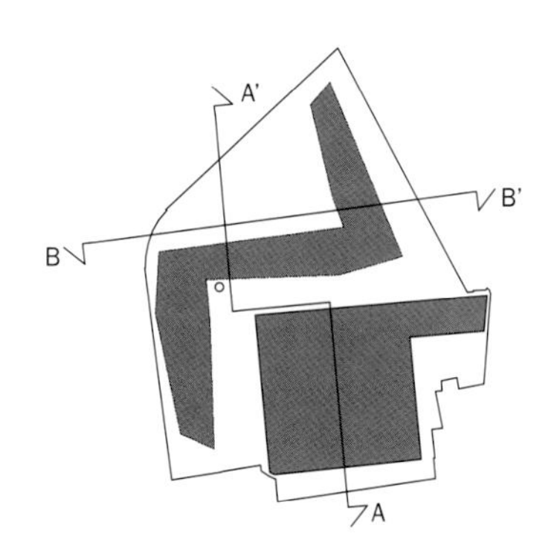

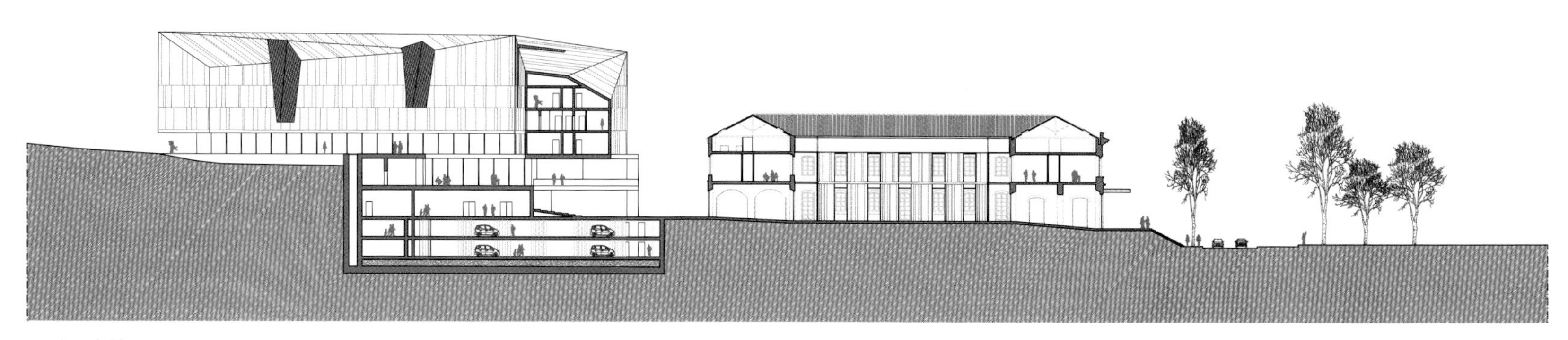

section A-A'
A-A'剖面

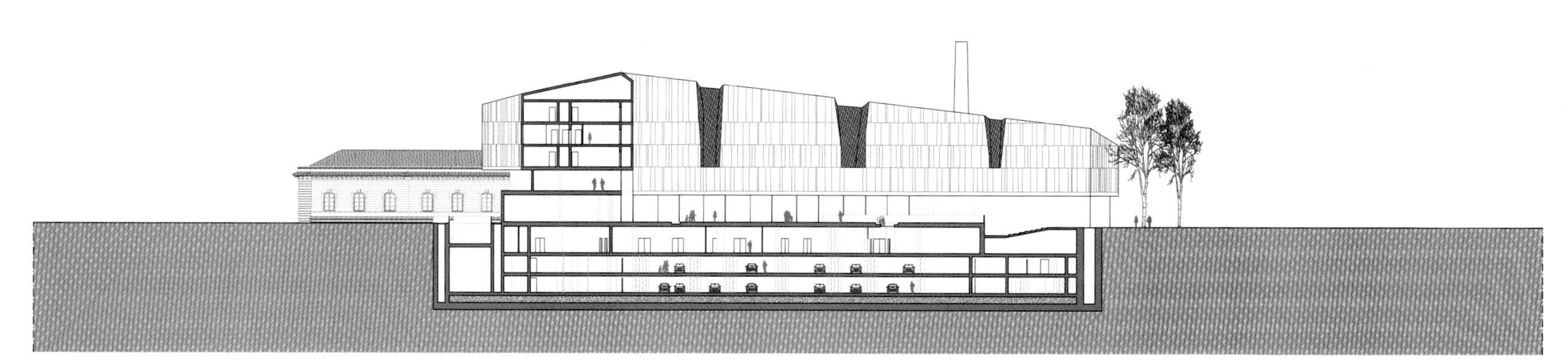

section B-B'
B-B'剖面

plan level +12.25
高度+12.25平面

PARK OF MUSIC AND CULTURE

Florence, Italy, 2007

[illegible] 伦萨，意大利
[illegible] 和礼堂
[illegible] Presidenza del Consiglio dei Ministri
[illegible] Favero&Milan Ingegneria
系统　Studio TI
规划　2007年设计竞赛，第2名
成本　800,000,000欧元
建筑面积　60,000平方米
承包商　Gia.Fi. Costruzioni S.p.a.

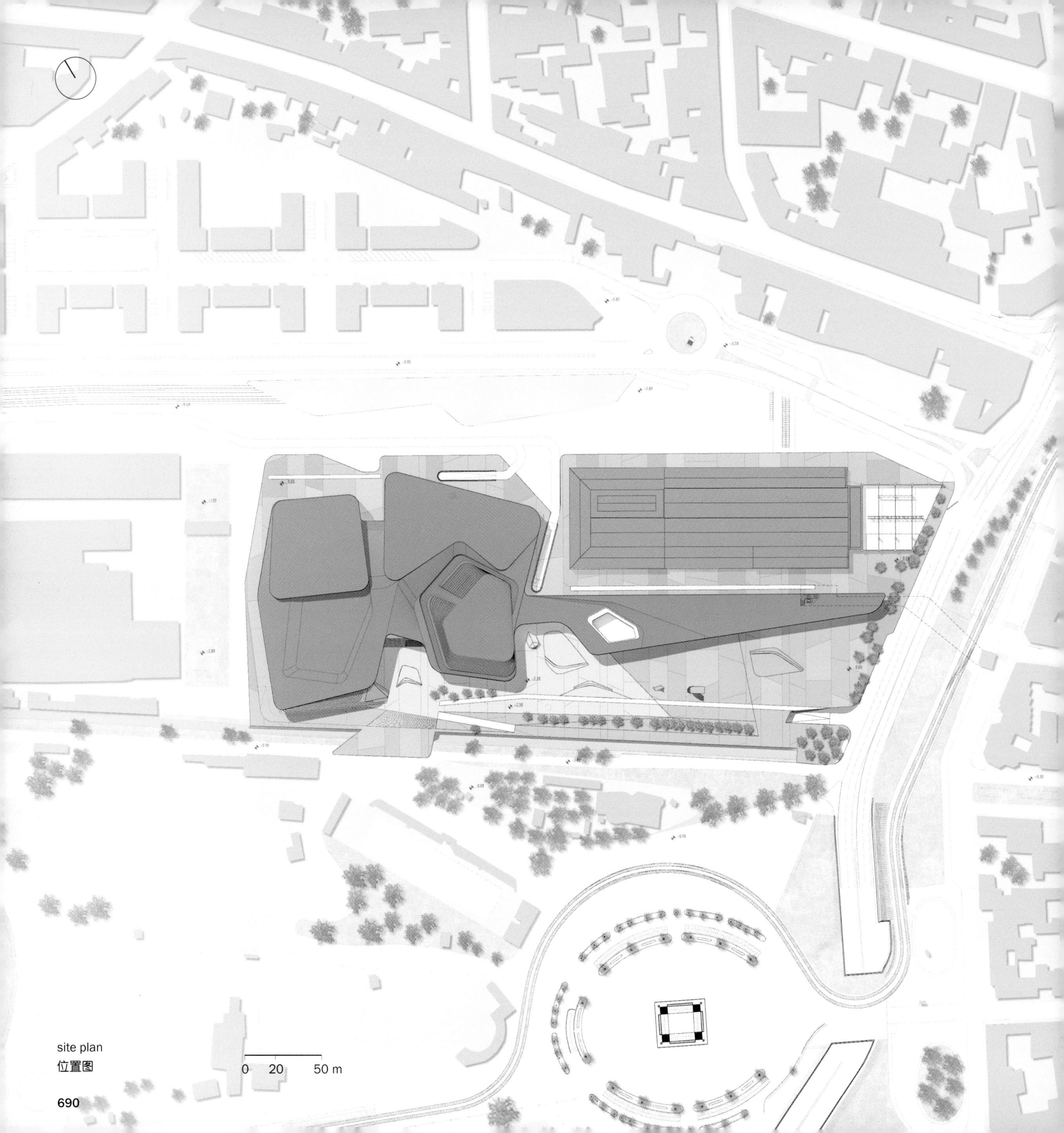

site plan
位置图

south-west elevation
西南正视图

north-east elevation
东北正视图

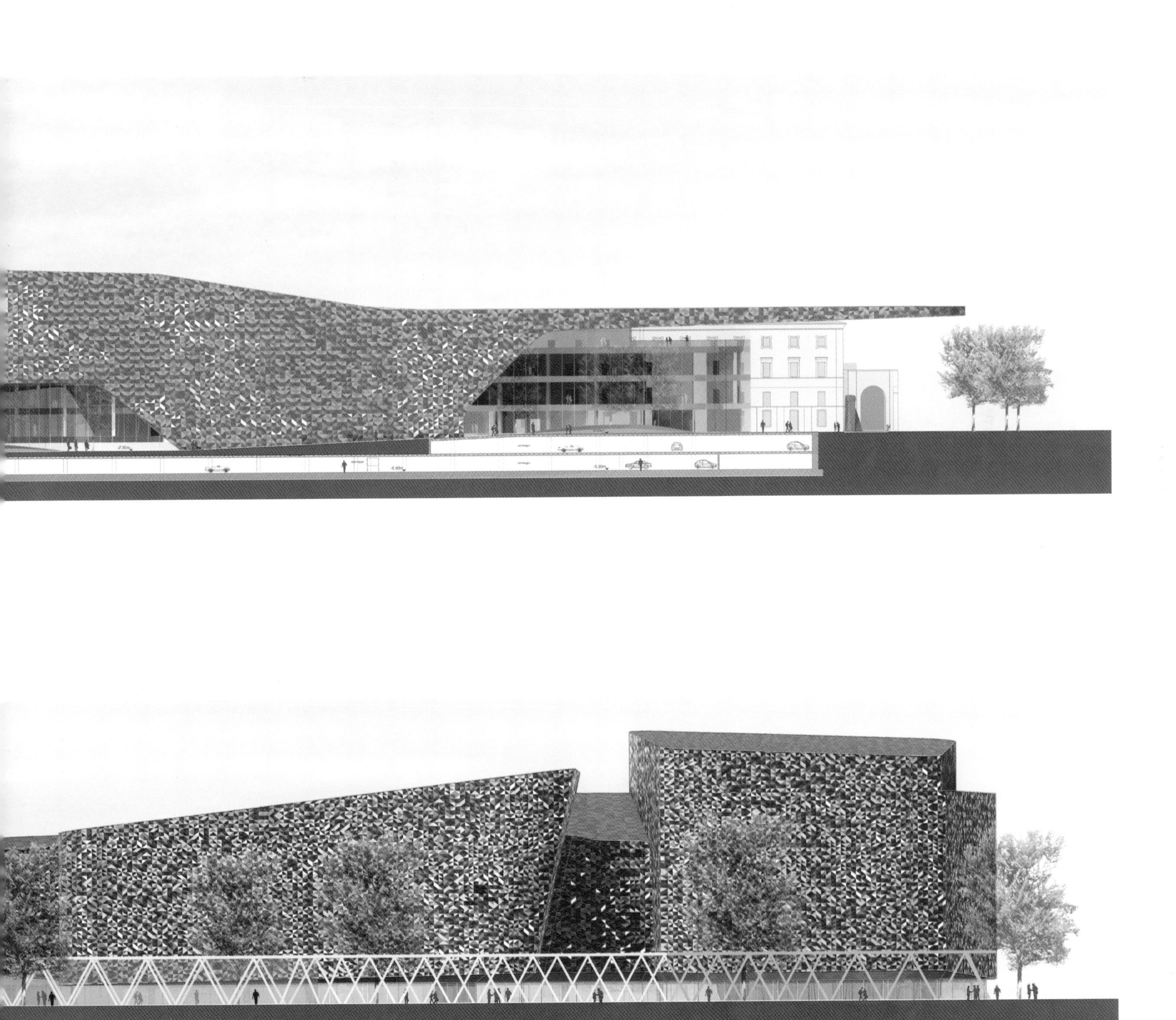

0 5 10 m

detail of the façade
正面细部

north-west elevation
西北正视图

south-east elevation
东南正视图

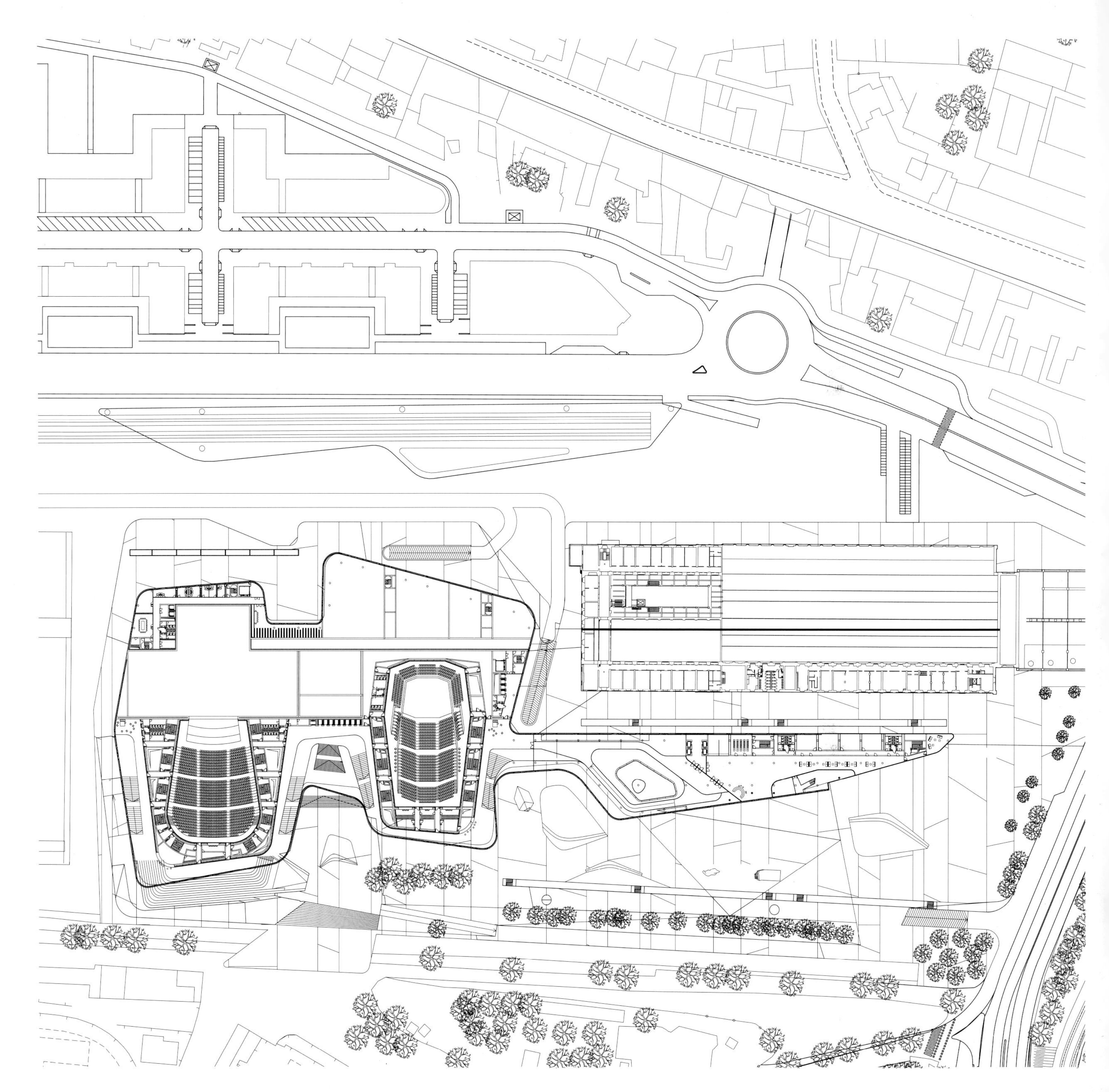

plan level +9.20

高度+9.20平面

0 10 20 m

section A-A'
A-A'剖面

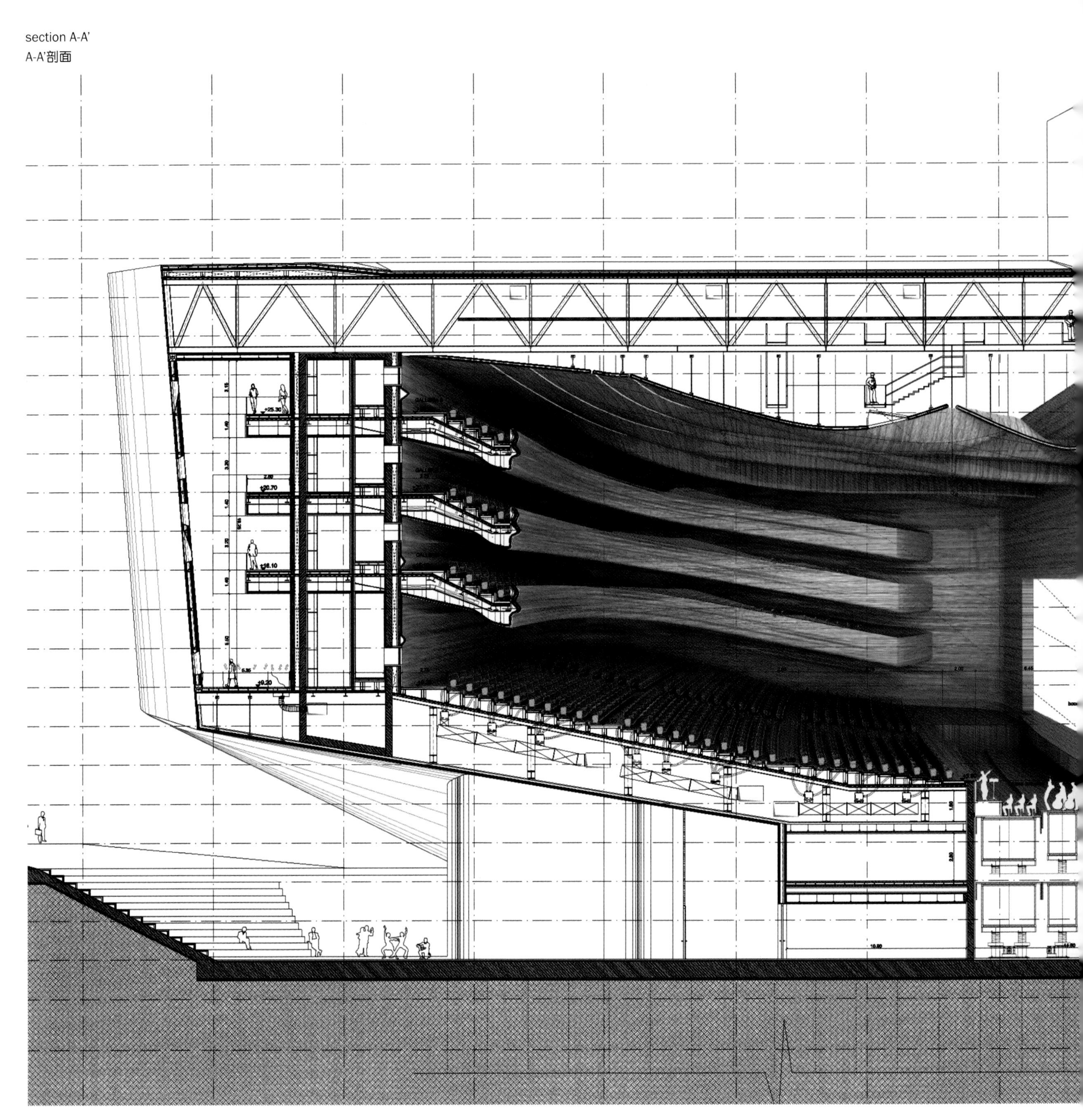

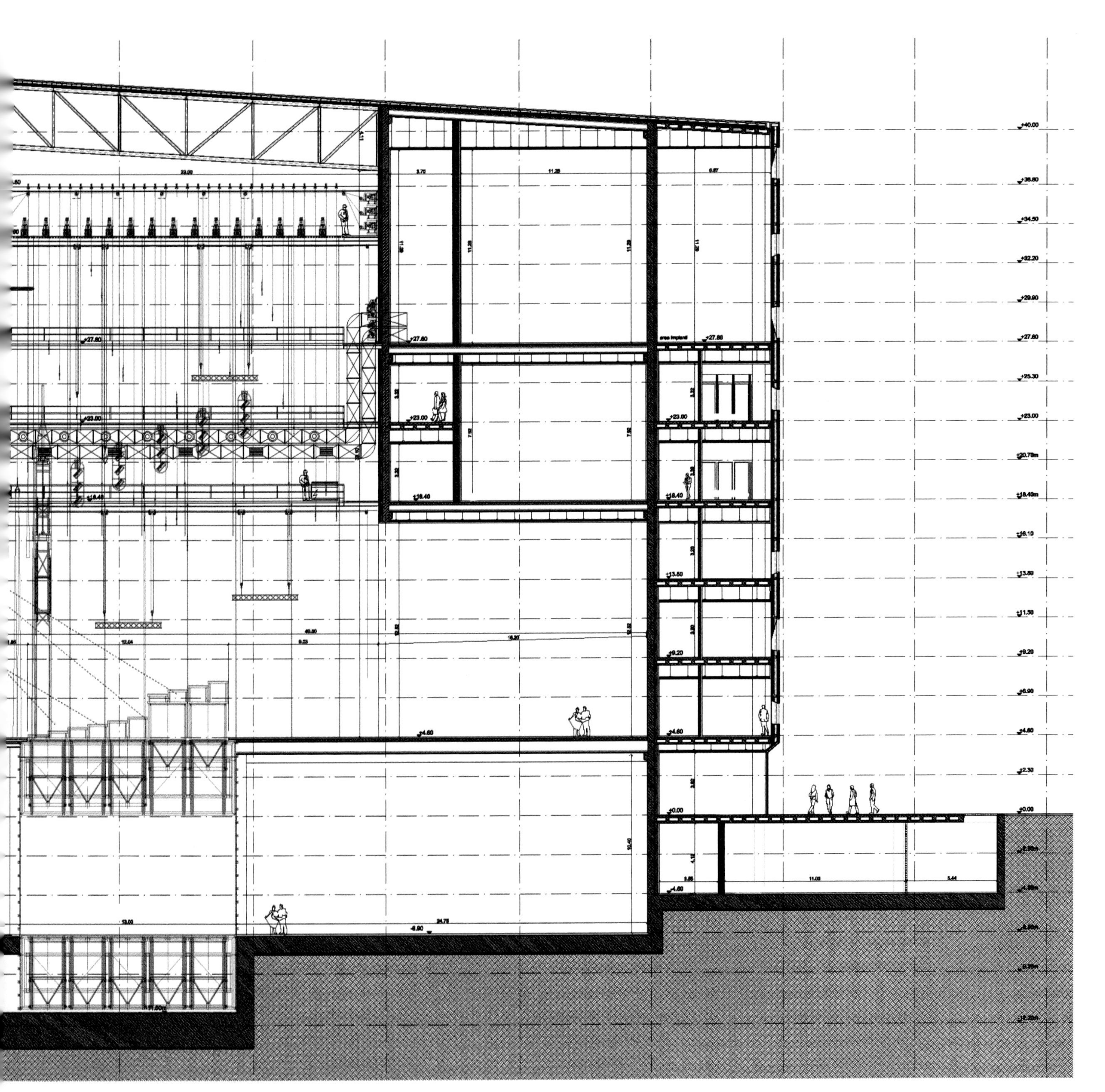

MILANOFIORI 2000

residential and commercial complex
Milan, Italy, 2007 - in progress

地点　Assago，米兰，意大利
项目　住宅和商业建筑
客户　Milanofiori 2000 S.r.l.
结构　Intertecno
系统　Intertecno
规划　2007年
建设　在建
成本　29,000,000欧元
建筑面积　24,848平方米

residential floor plan, second level
居住层规划，第二层

0 10 m

offices floor plan, third level

办公层规划，第三层

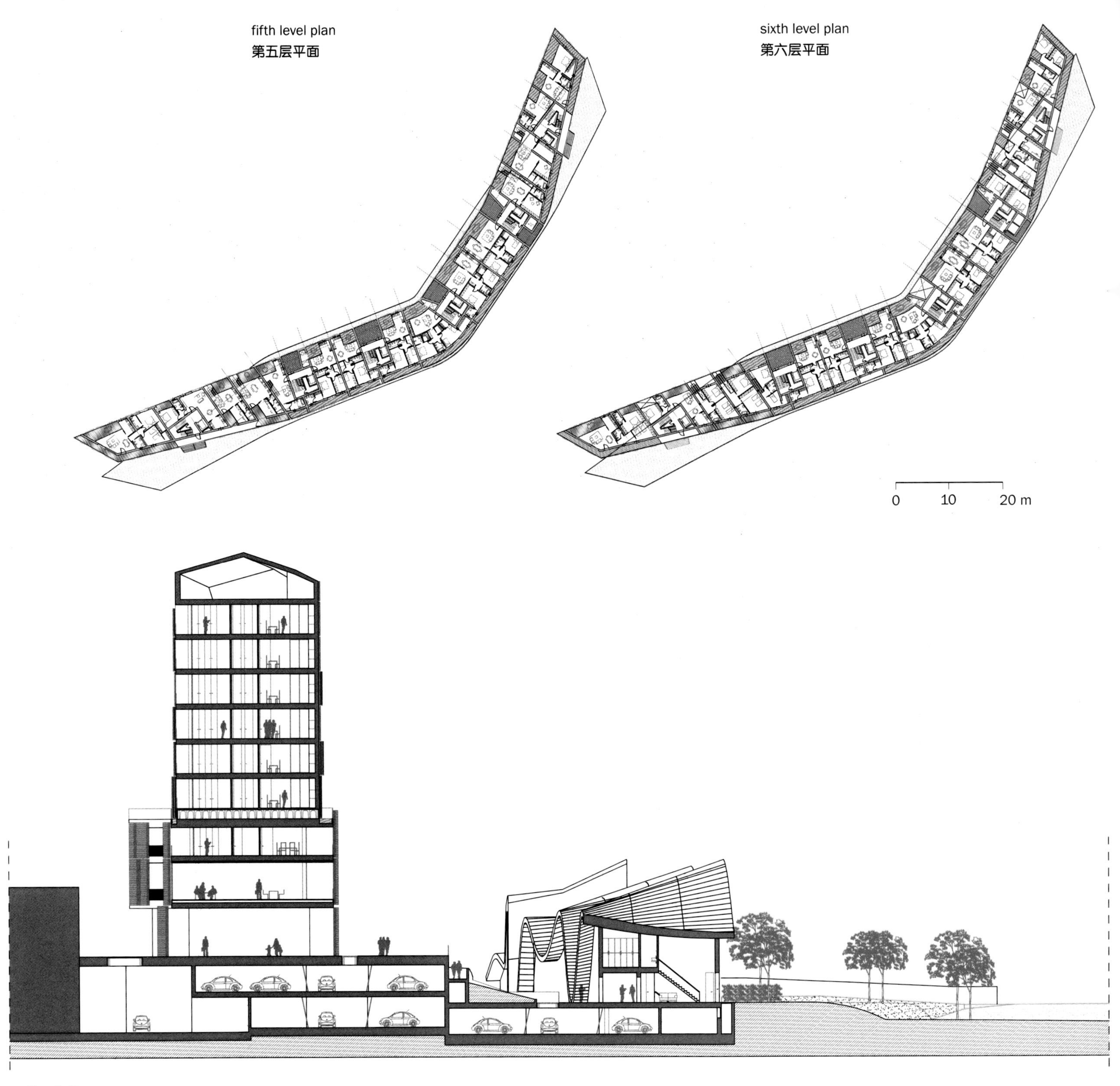
fifth level plan
第五层平面
sixth level plan
第六层平面
0
10
20 m
section A-A’
A-A’剖面

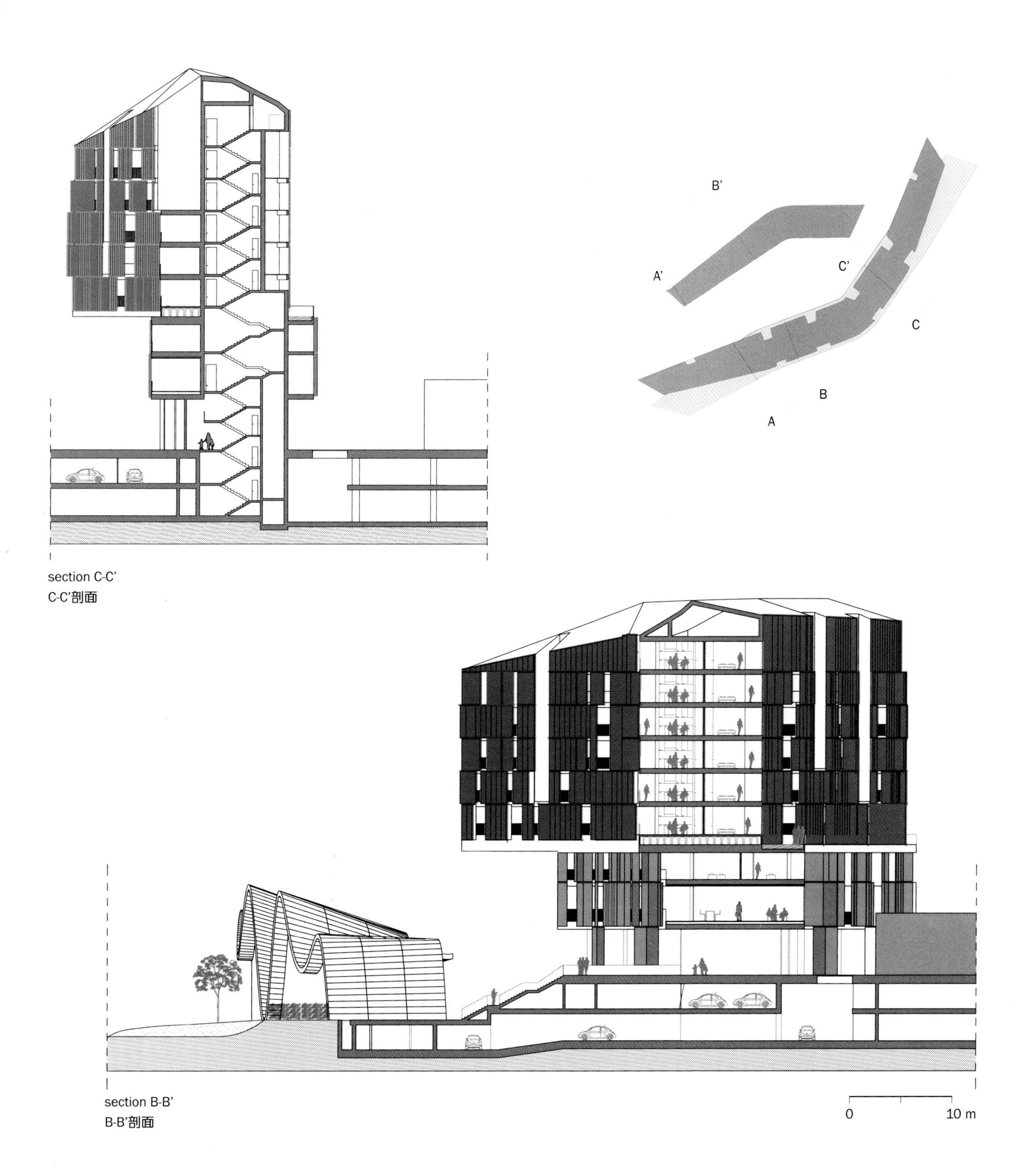

section C-C'
C-C'剖面

section B-B'
B-B'剖面

GEL - GREEN ENERGY LABORATORY

Shanghai, China, 2008 - in progress

地点 上海交通大学闵行校区，上海，中国
项目 研究中心
客户 上海交通大学
结构 Favero&Milan Ingegneria
系统 TIFS Ingegneria
规划 2008年
建设 在建
占地面积 1,500平方米
建筑面积 4,850平方米
体量 27,000立方米

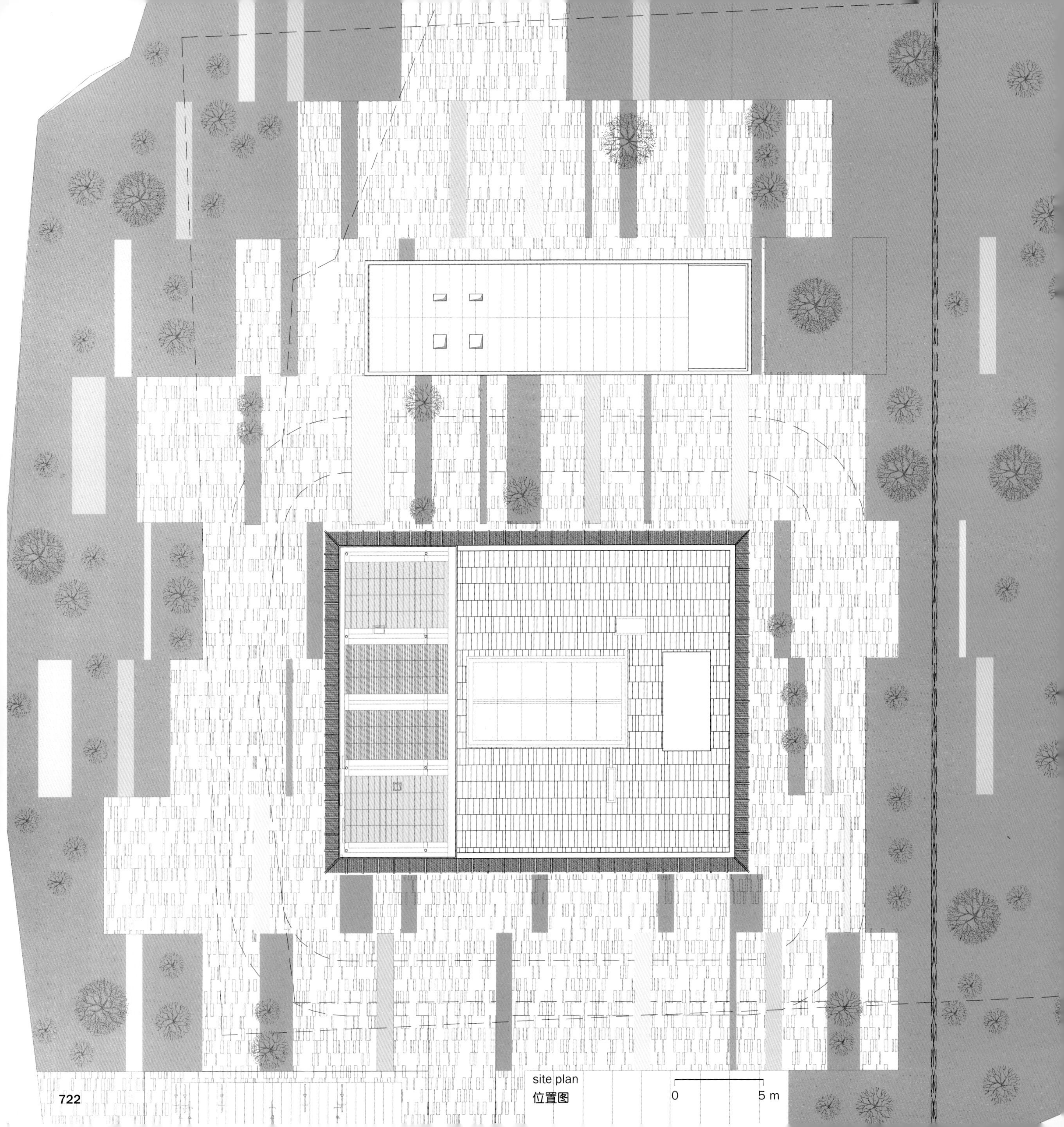

site plan
位置图

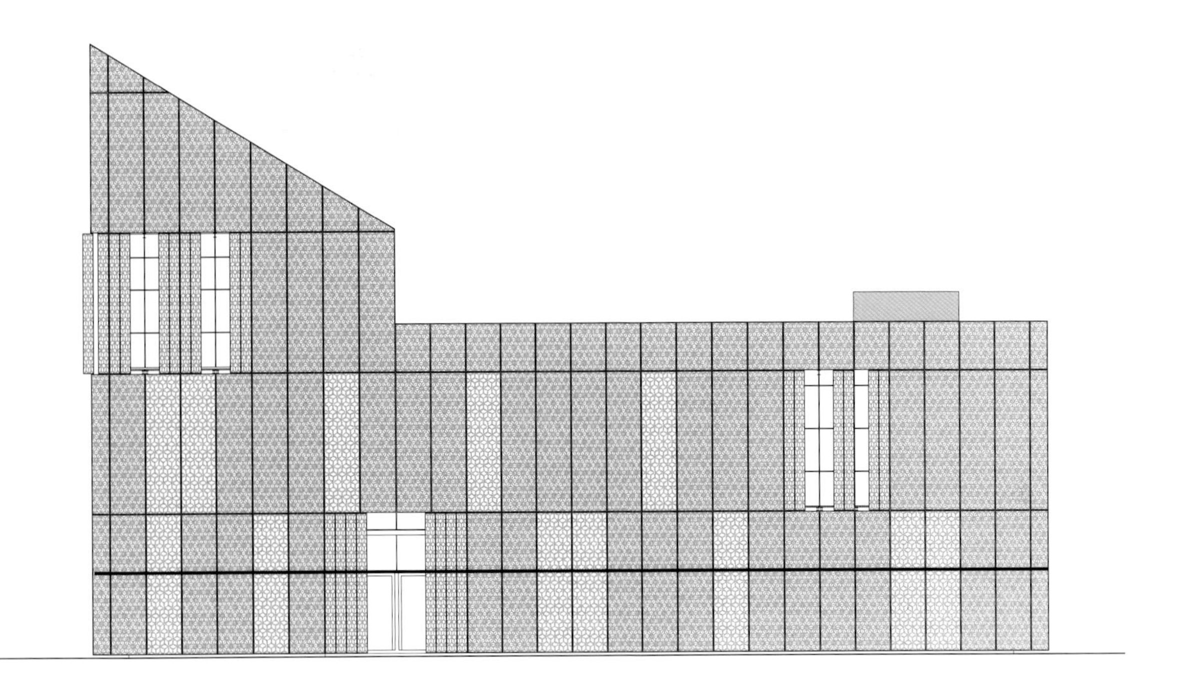

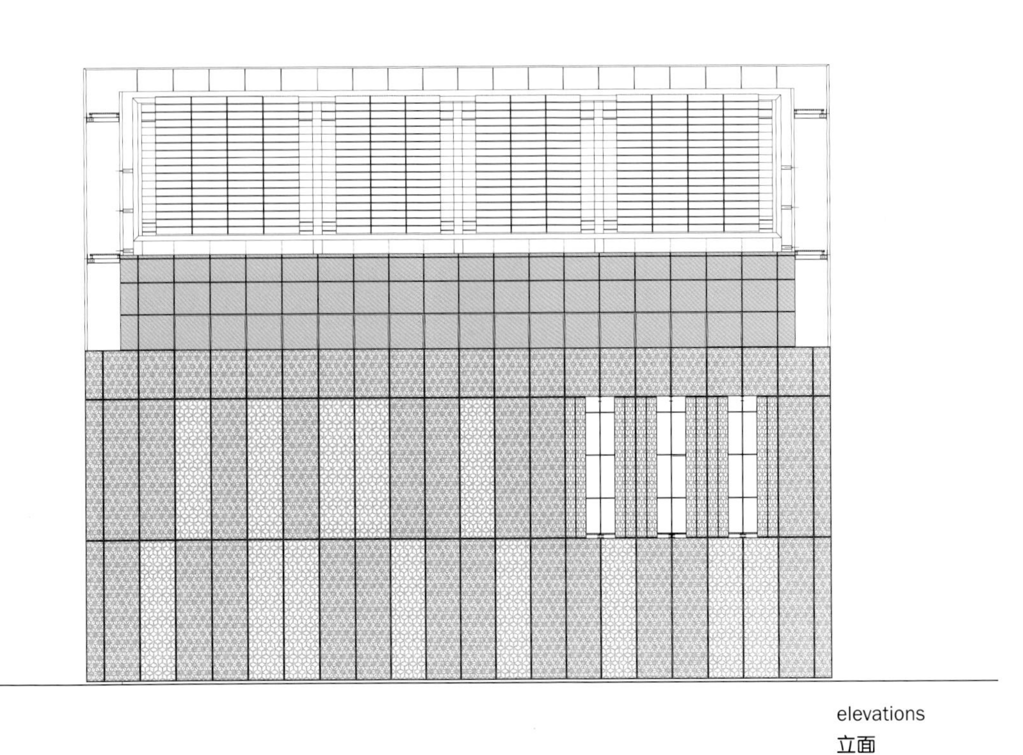

elevations
立面

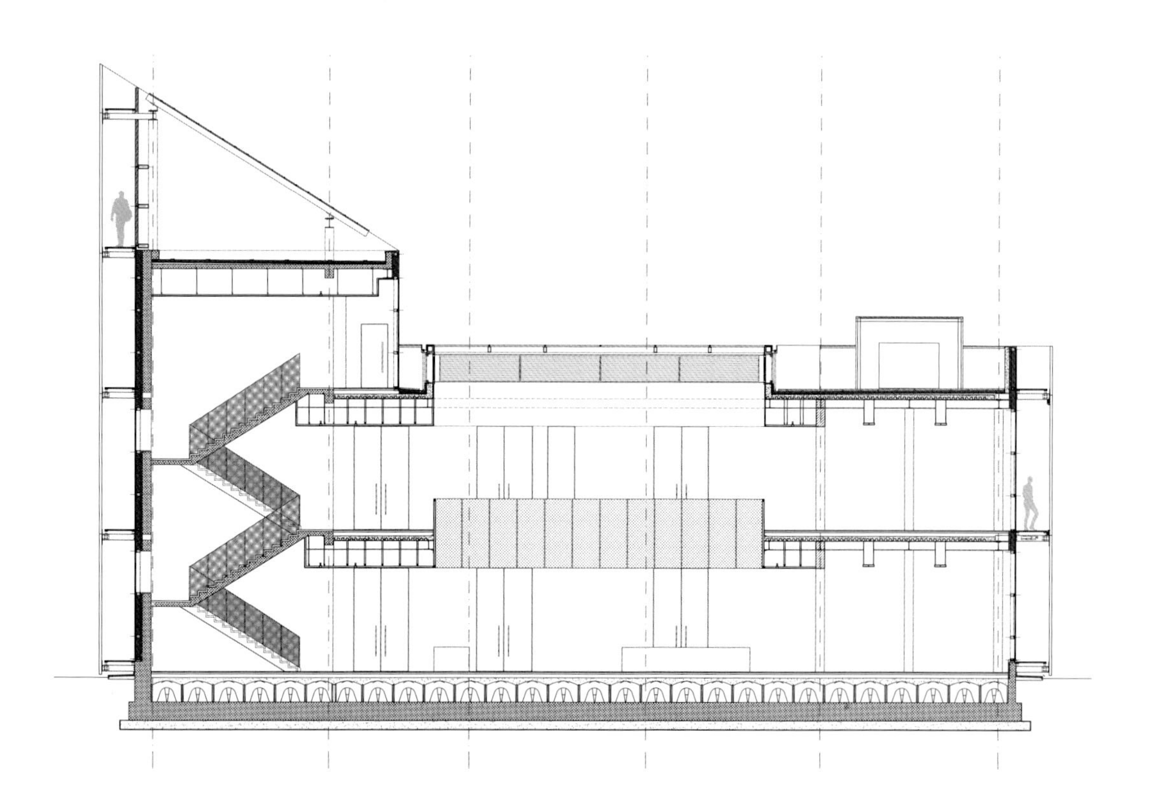

longitudinal section
纵剖面

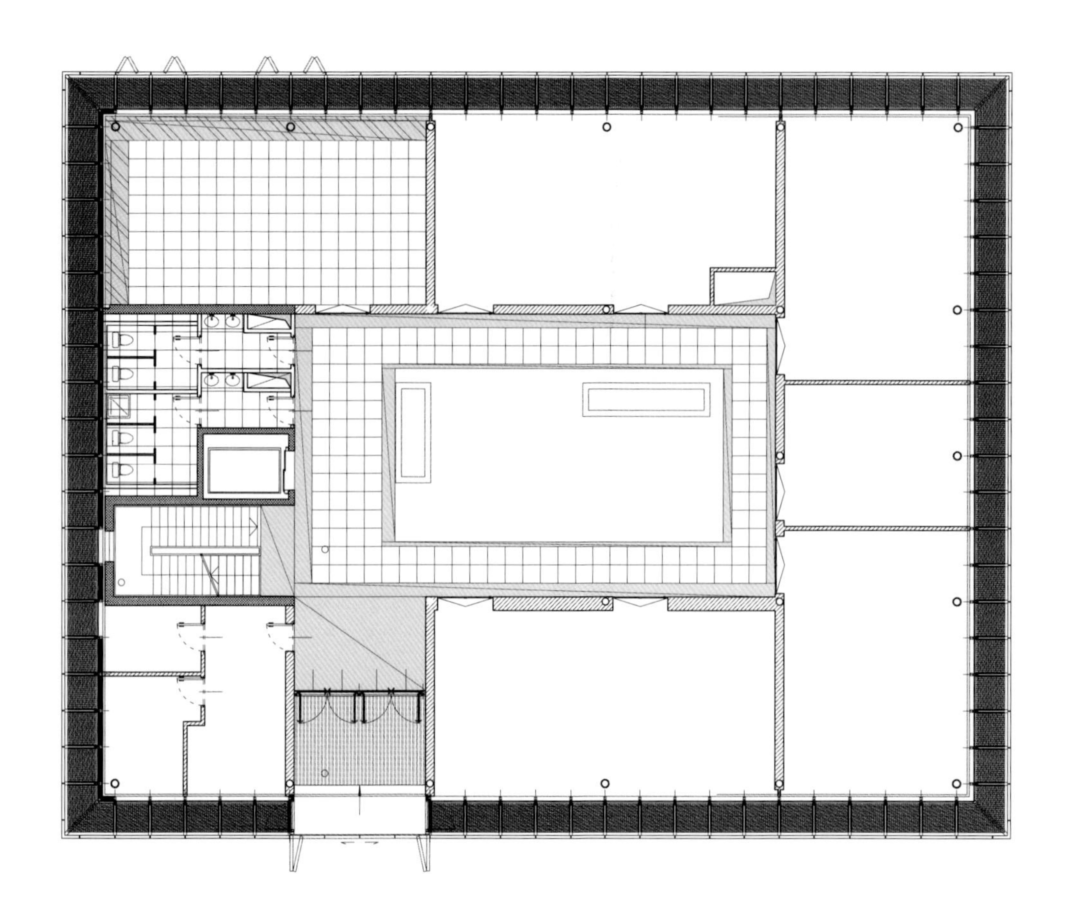

ground floor plan
一层平面

0 5 10 m

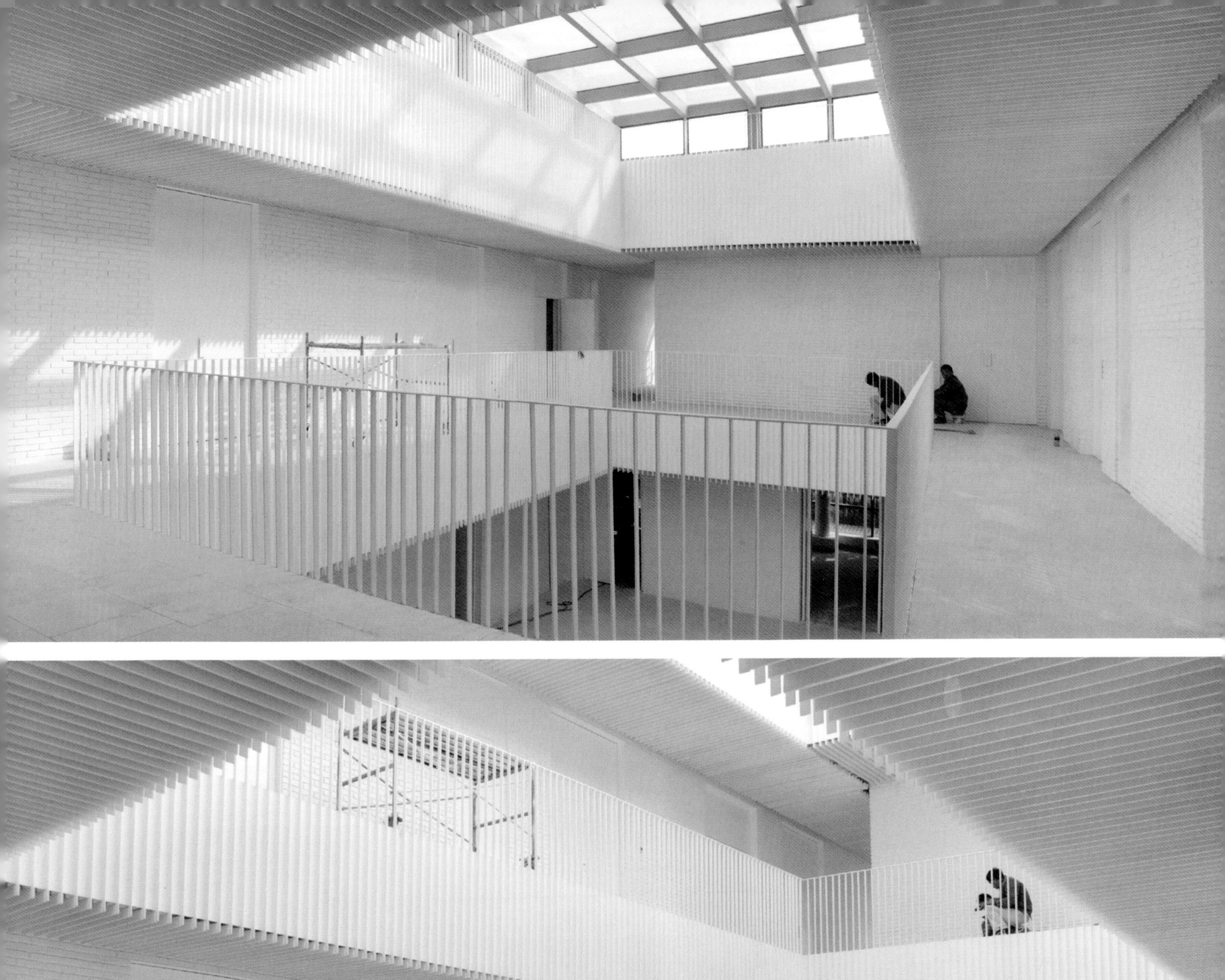

ARCHITECTURE DEPARTMENT BUILDING

Tripoli, Libya, 2008

地点 的黎波里，利比亚
项目 公共建筑
客户 Odac – Meftah Waggah
当地咨询公司 N.C.B., Mustafà Mezughi, Mohamed Gheblawi
规划 2008
成本 15,000,000欧元
建筑面积 8,000平方米
体量 30,000立方米

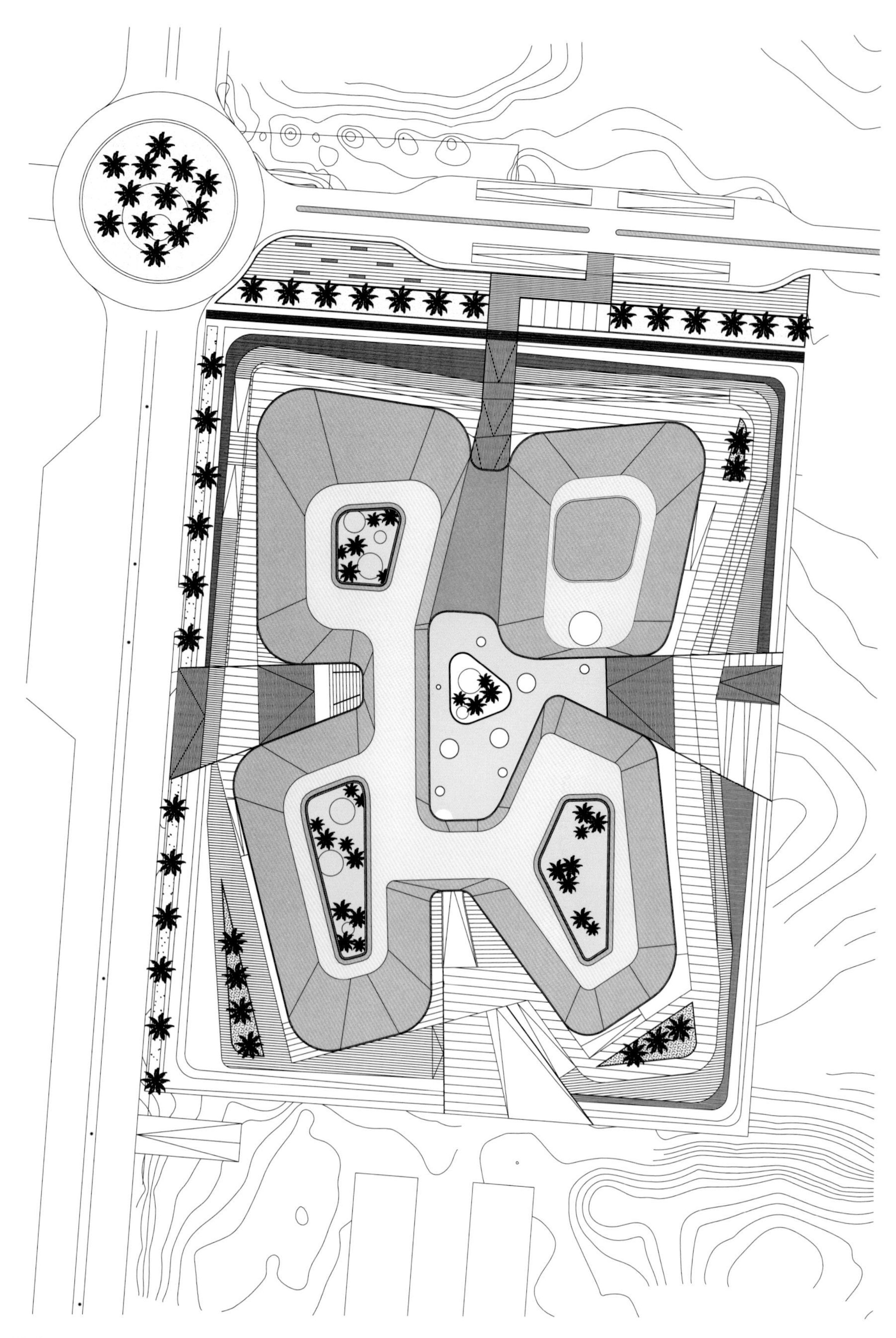

roofing
顶部

0 10 20 m

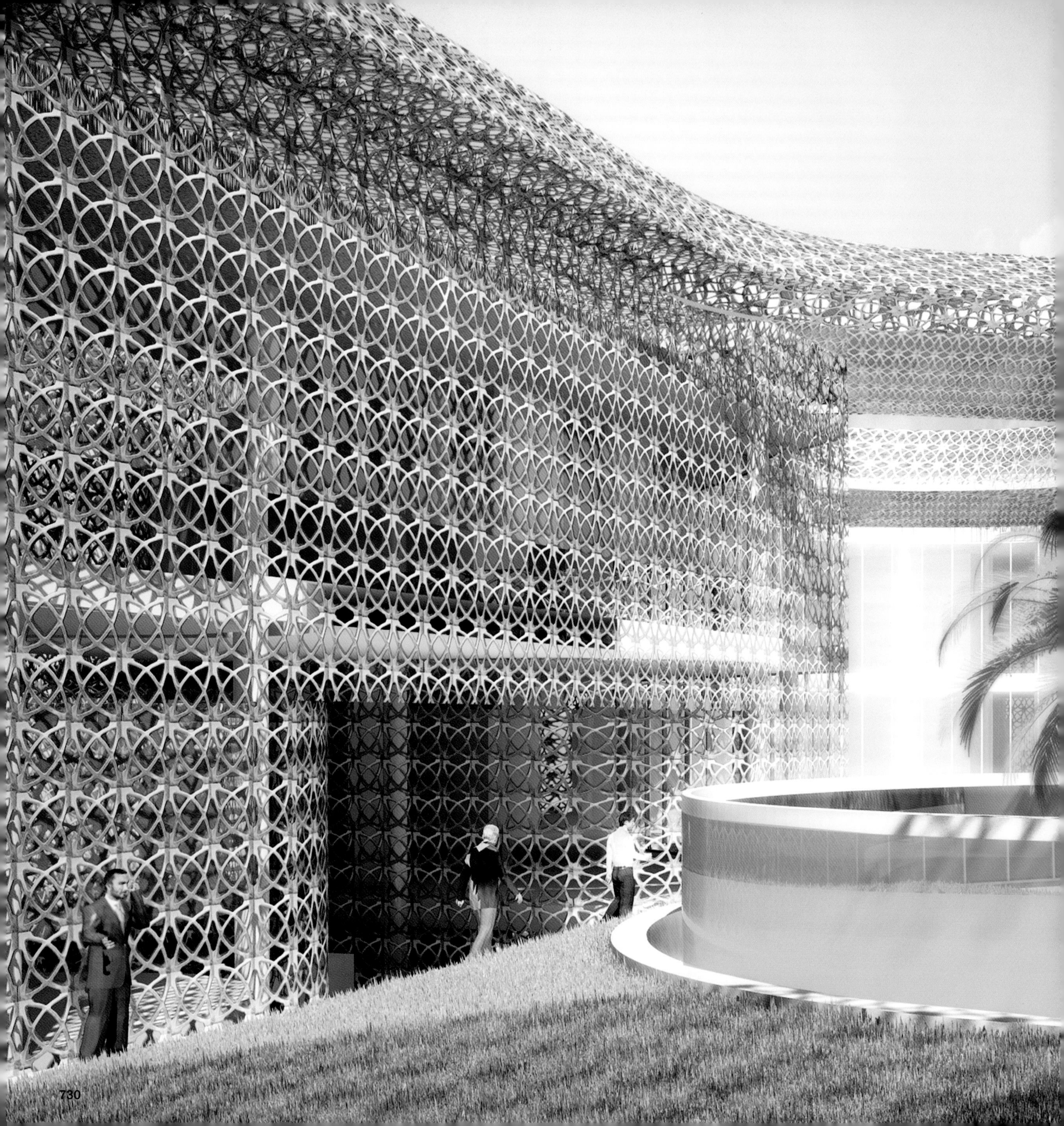

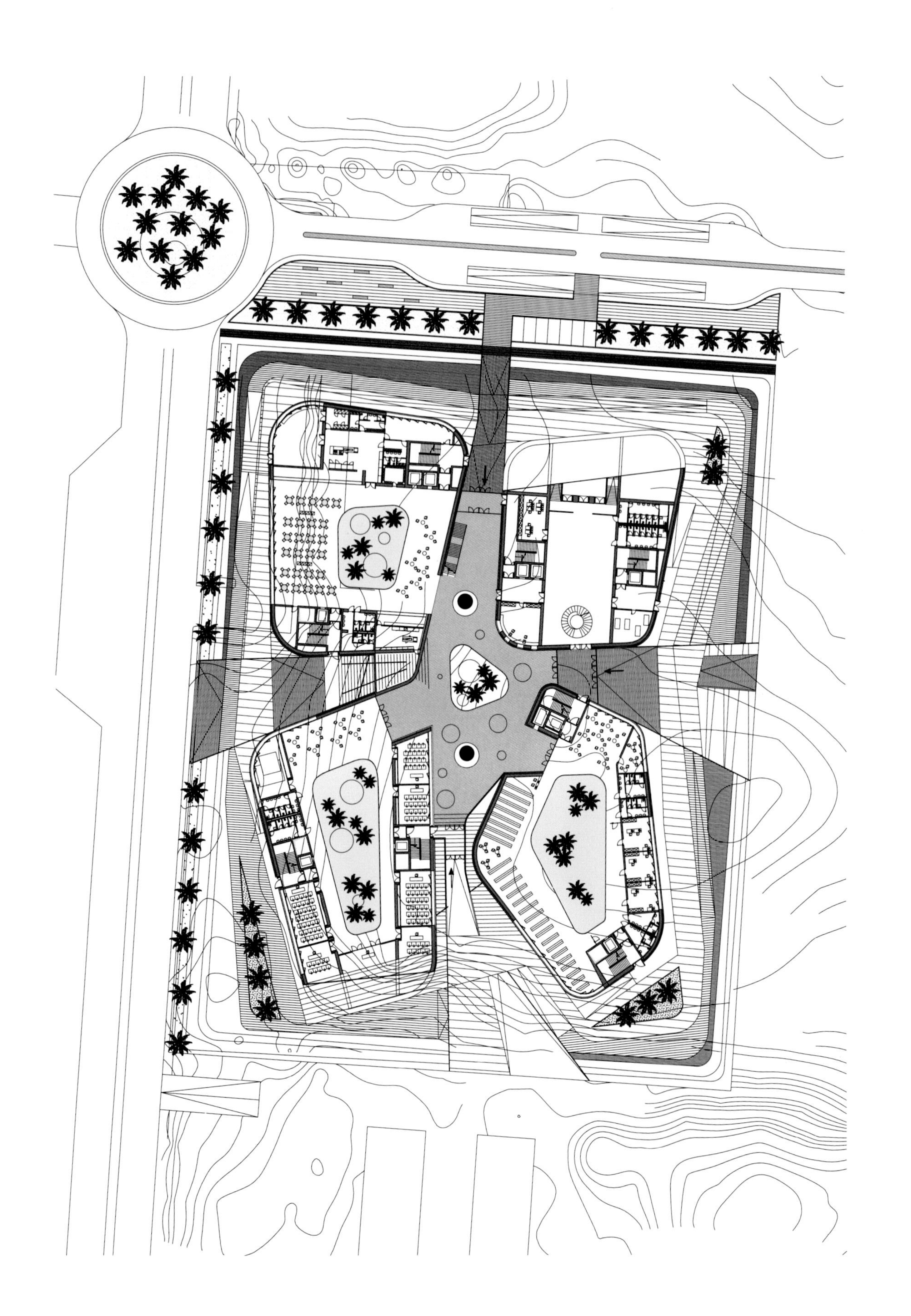

ground floor plan

一层平面

0 10 20 m

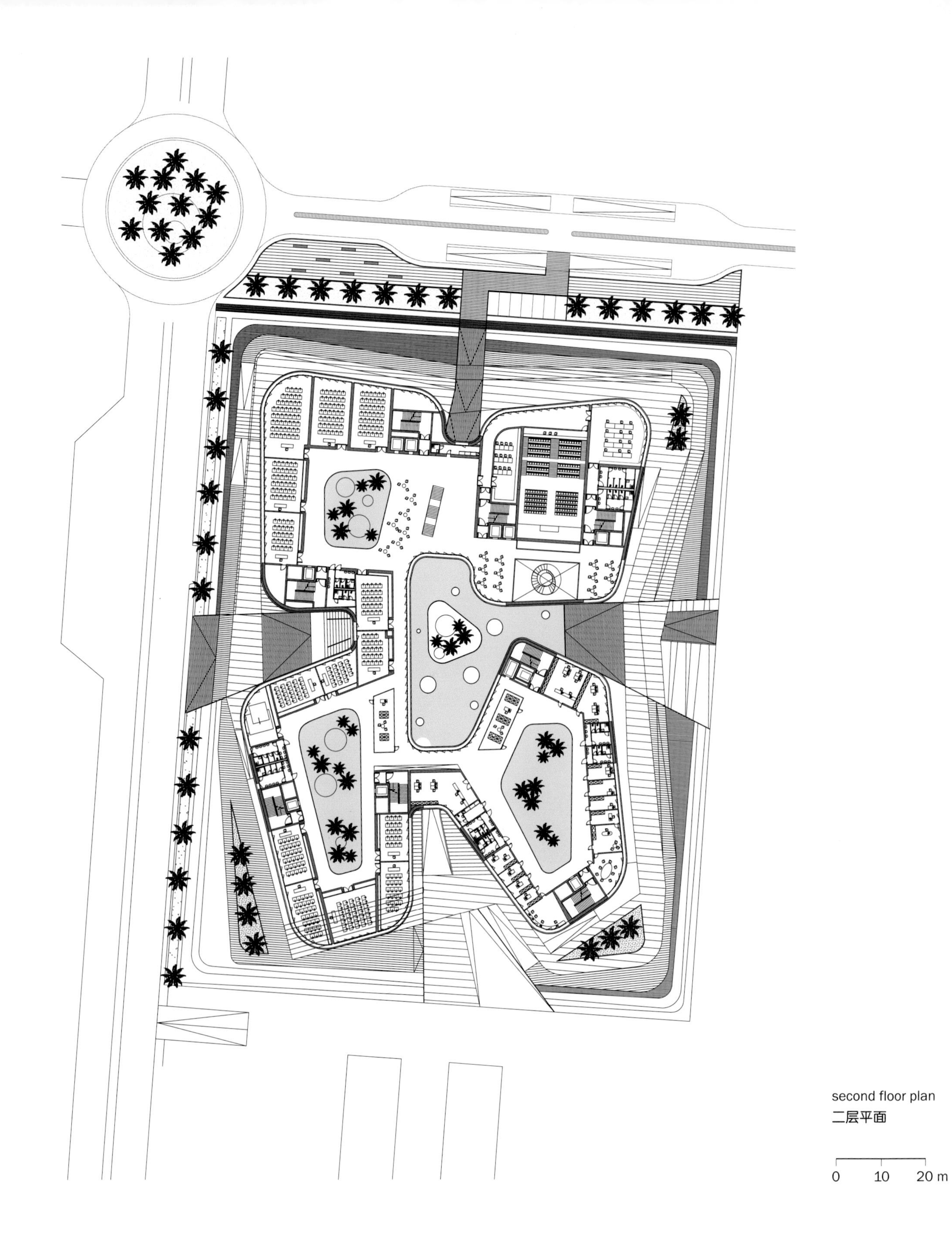

second floor plan
二层平面

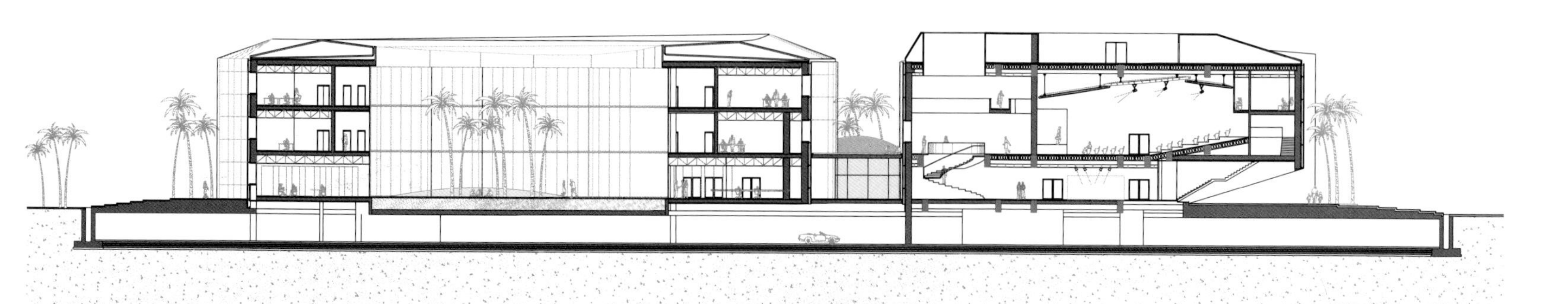

longitudinal section
纵剖面

0 5 10 m

KPM TOWER

residential and commercial complex
Dubai, United Arab Emirates, 2009-2010

地点　杜拜，阿拉伯联合酋长国
项目　住宅和商业建筑群
客户　Marina Exclusive L.t.d.
结构　Sinergo
结构咨询机构　aei progetti - Niccolò De Robertis
系统　Sinergo
规划　2009年-2010年
成本　60,000,000欧元
占地面积　3,382平方米
建筑面积　29,800平方米
体量　110,000立方米
承包商　GTCC

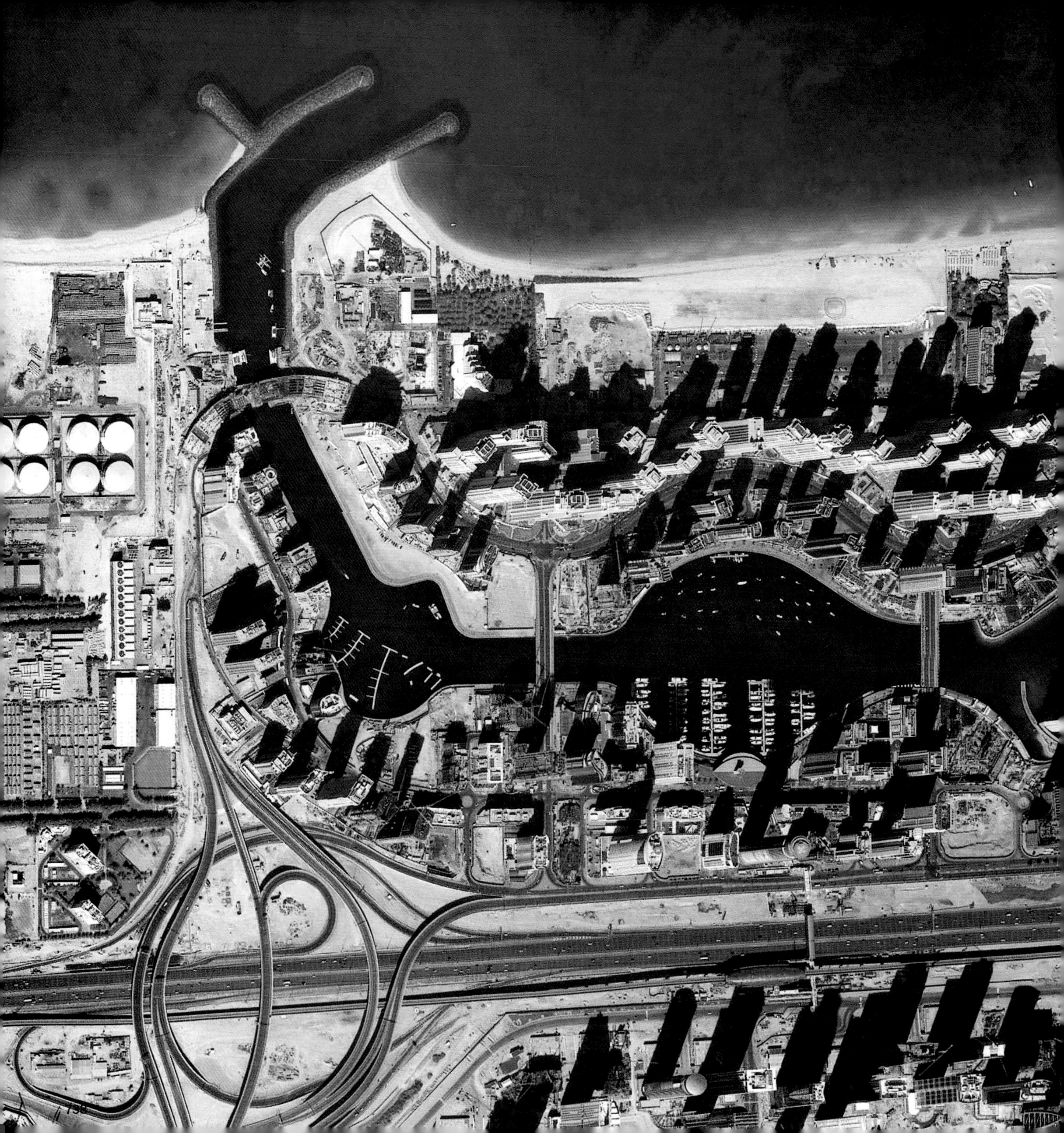

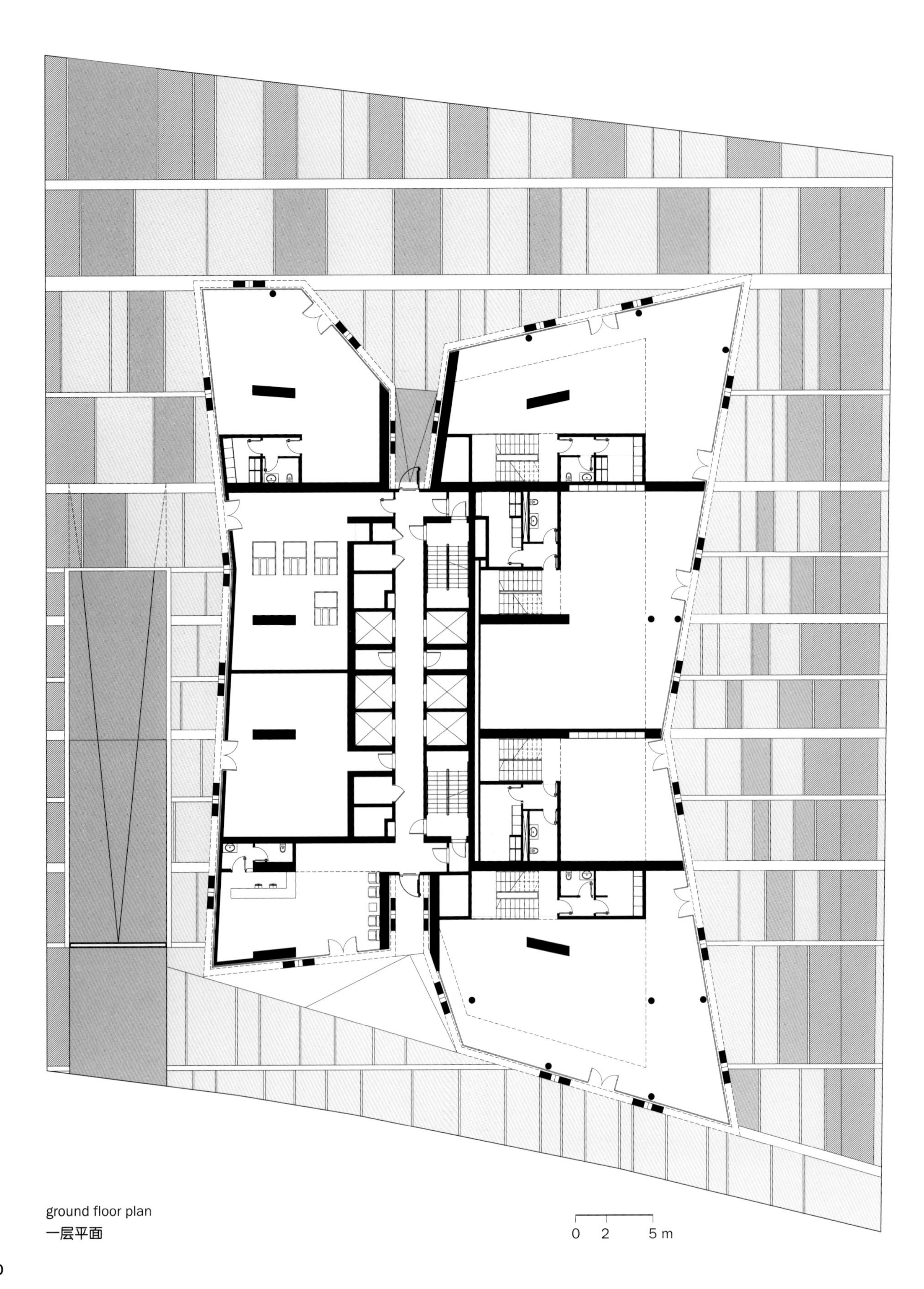

ground floor plan
一层平面

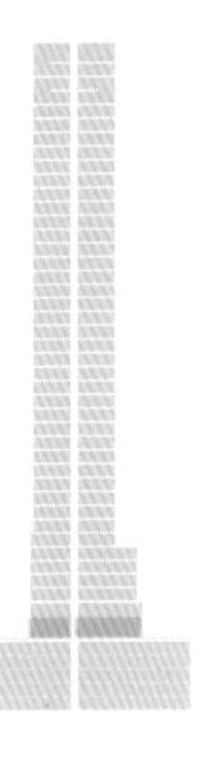

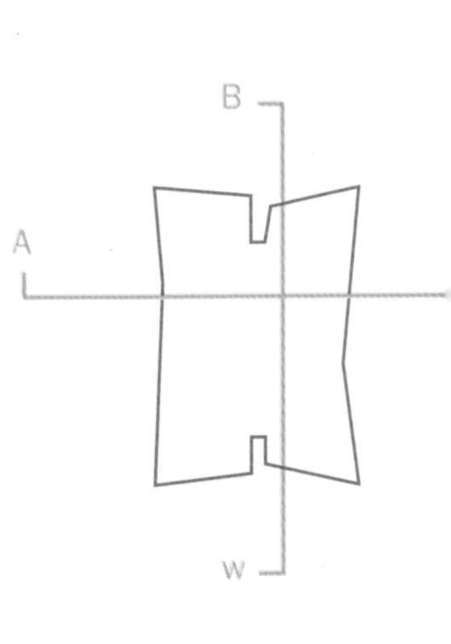
B
A
W

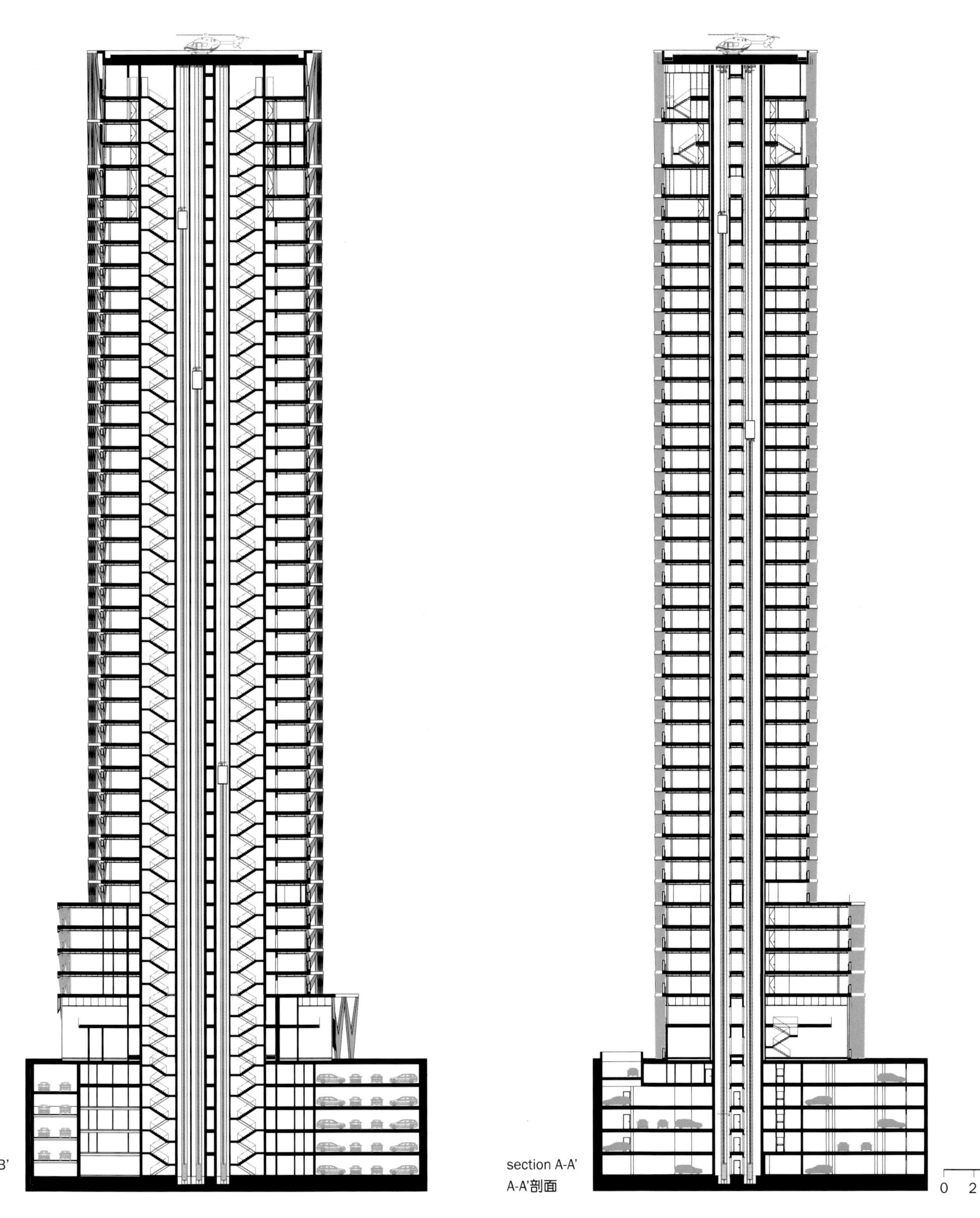

section B-B'
B-B'剖面

section A-A'
A-A'剖面

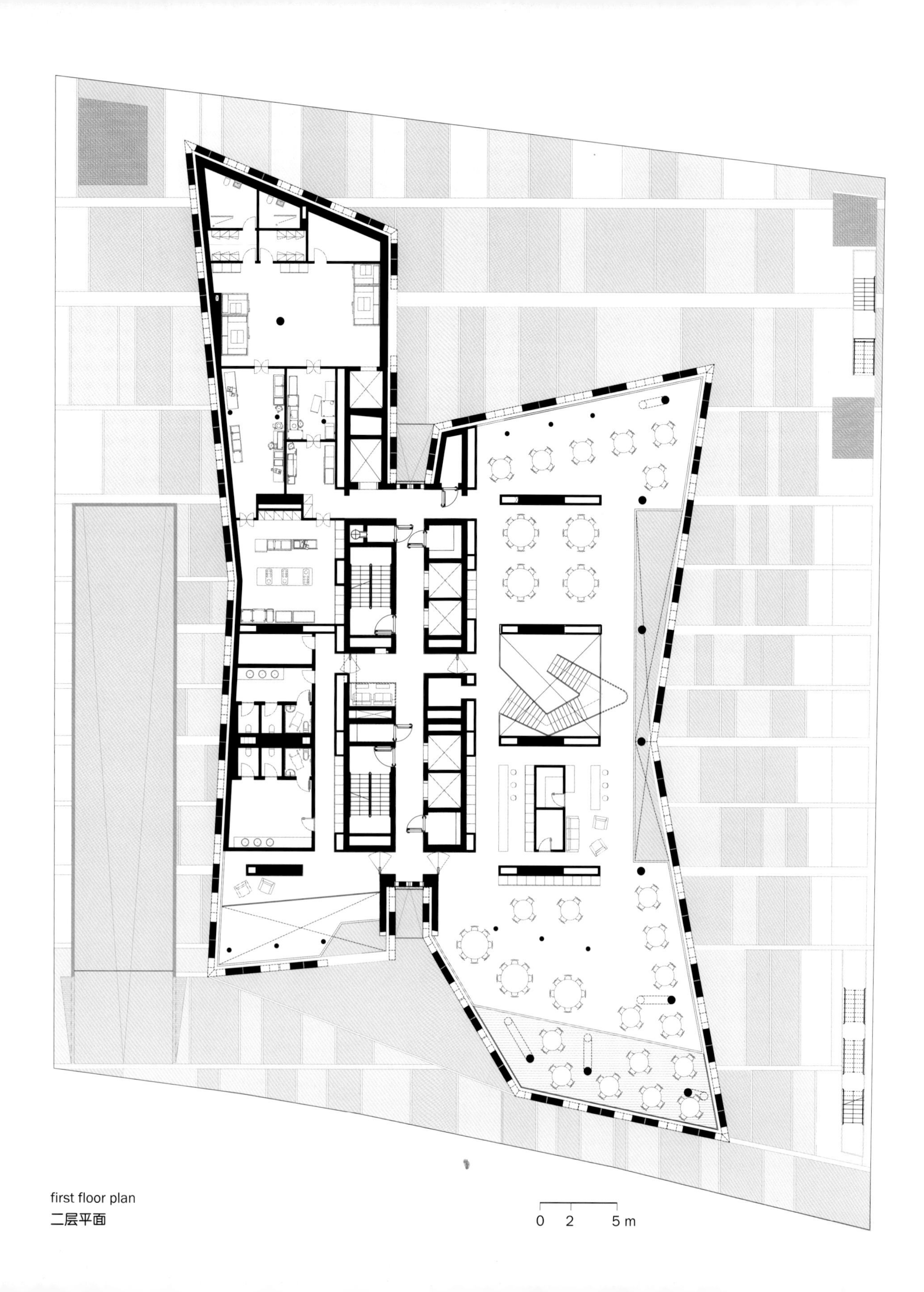

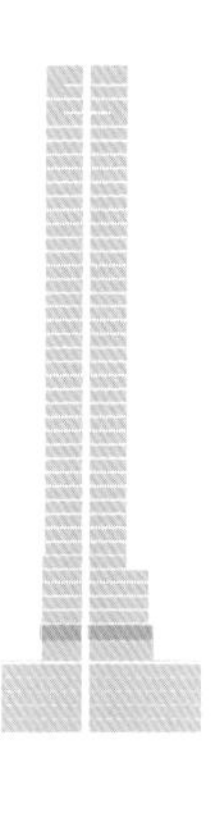

first floor plan
二层平面

health club plan
健身中心平面

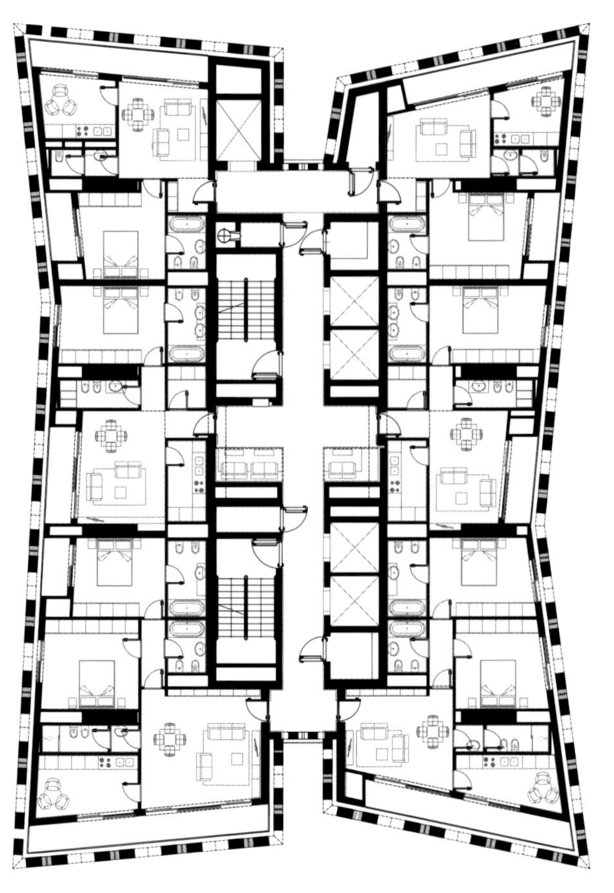

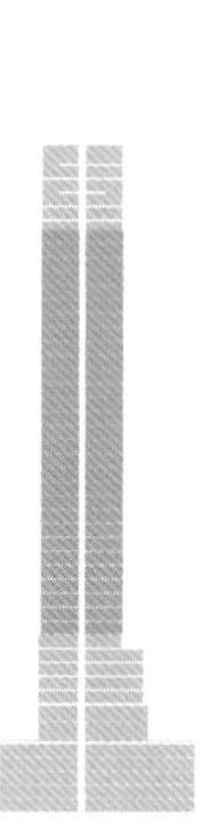

plan from 7th to 35th level
七至三十五层平面

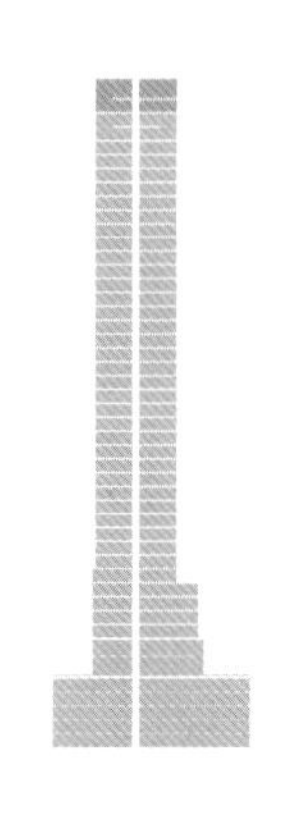

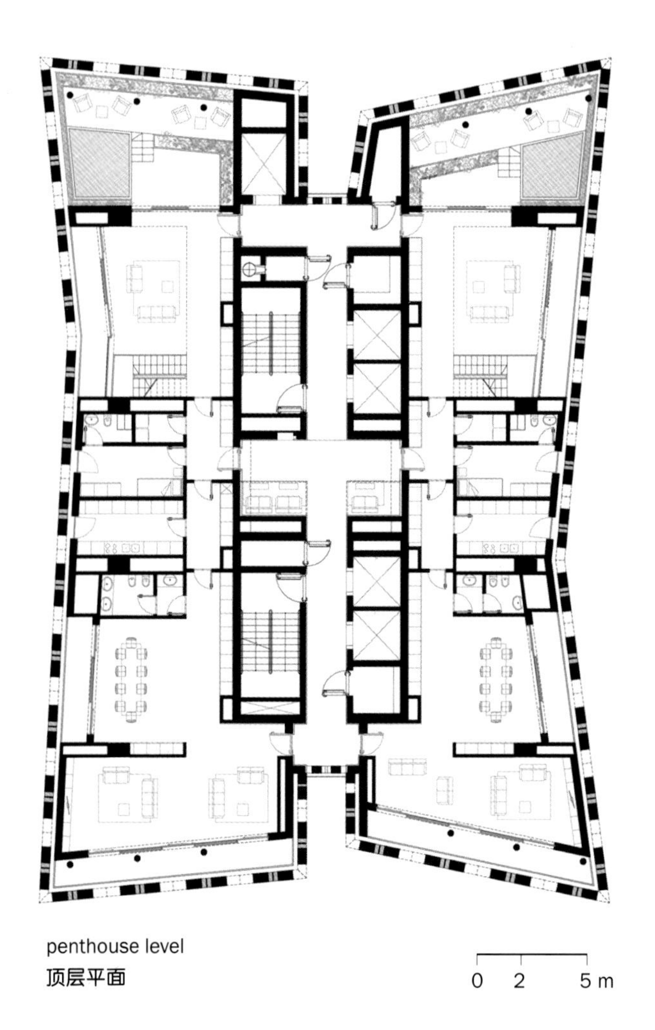

penthouse level
顶层平面

ART CUBE

Casabeltrame, Novara, Italy, 2009

地点 Casalbeltrame，诺瓦腊，意大利
项目 雕塑展厅
客户 Materima S.r.l.
结构 Map Carpenteria Metallica S.r.l.
规划 2009年
建设 2009年
成本 157,000欧元
表面积 81平方米

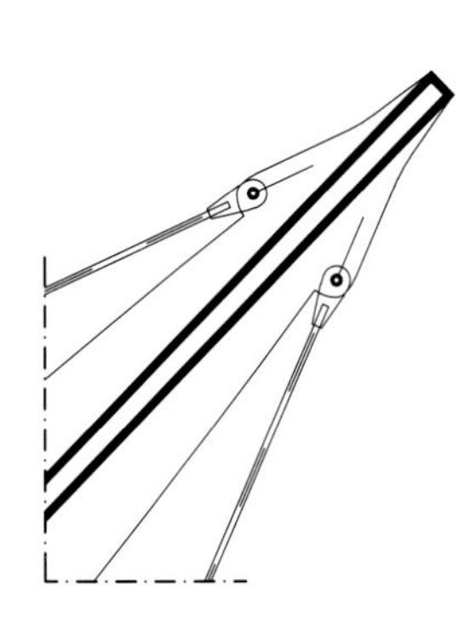

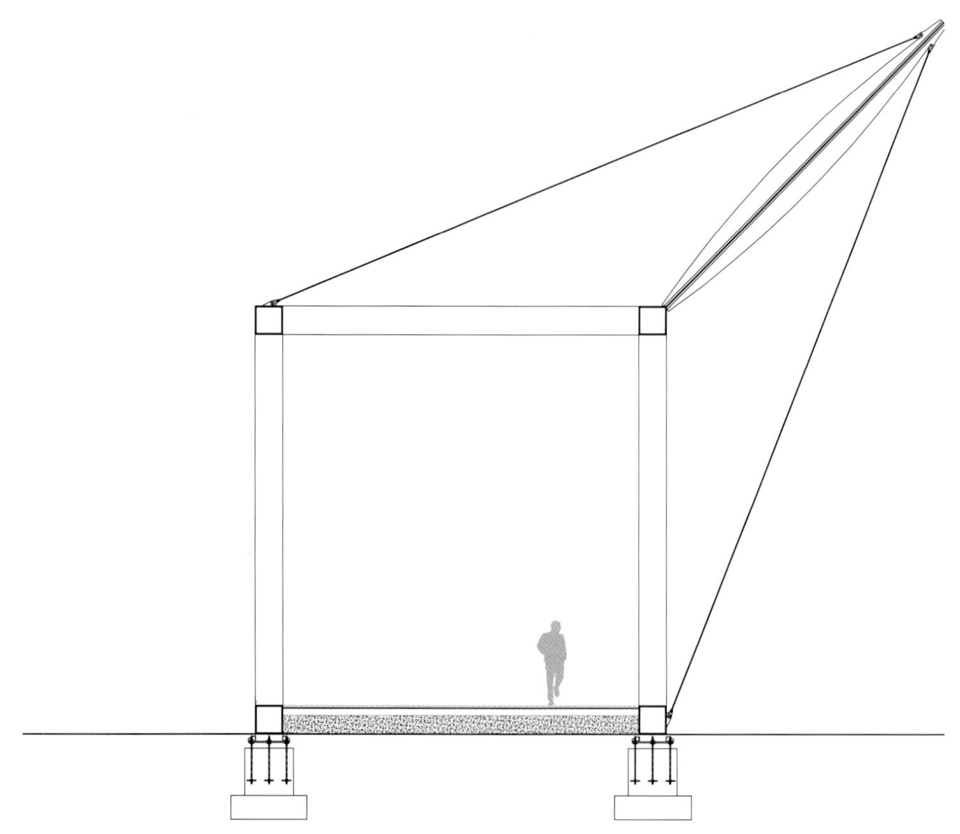

section
剖面

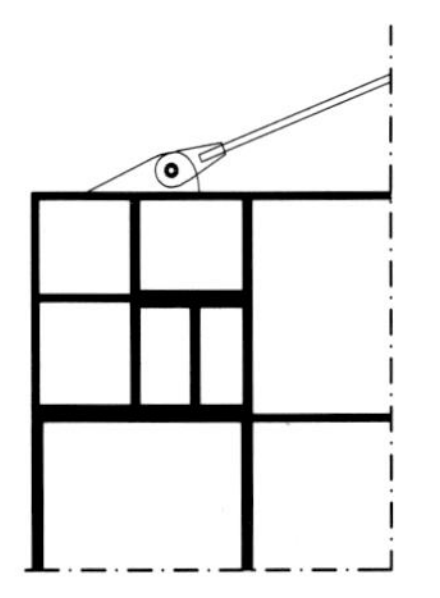

strut details
支柱细部

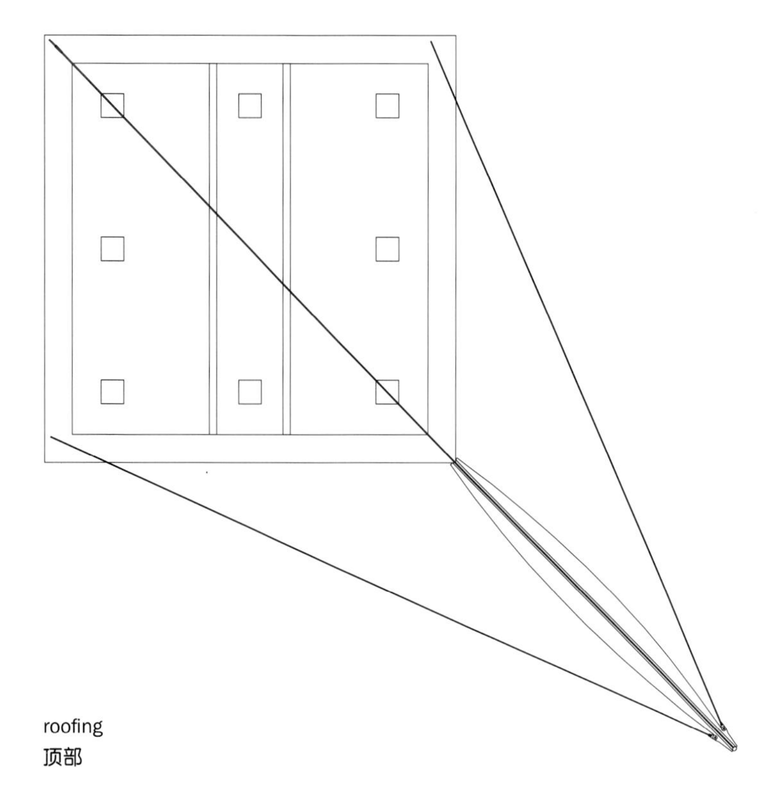

roofing
顶部

0 1 2 m

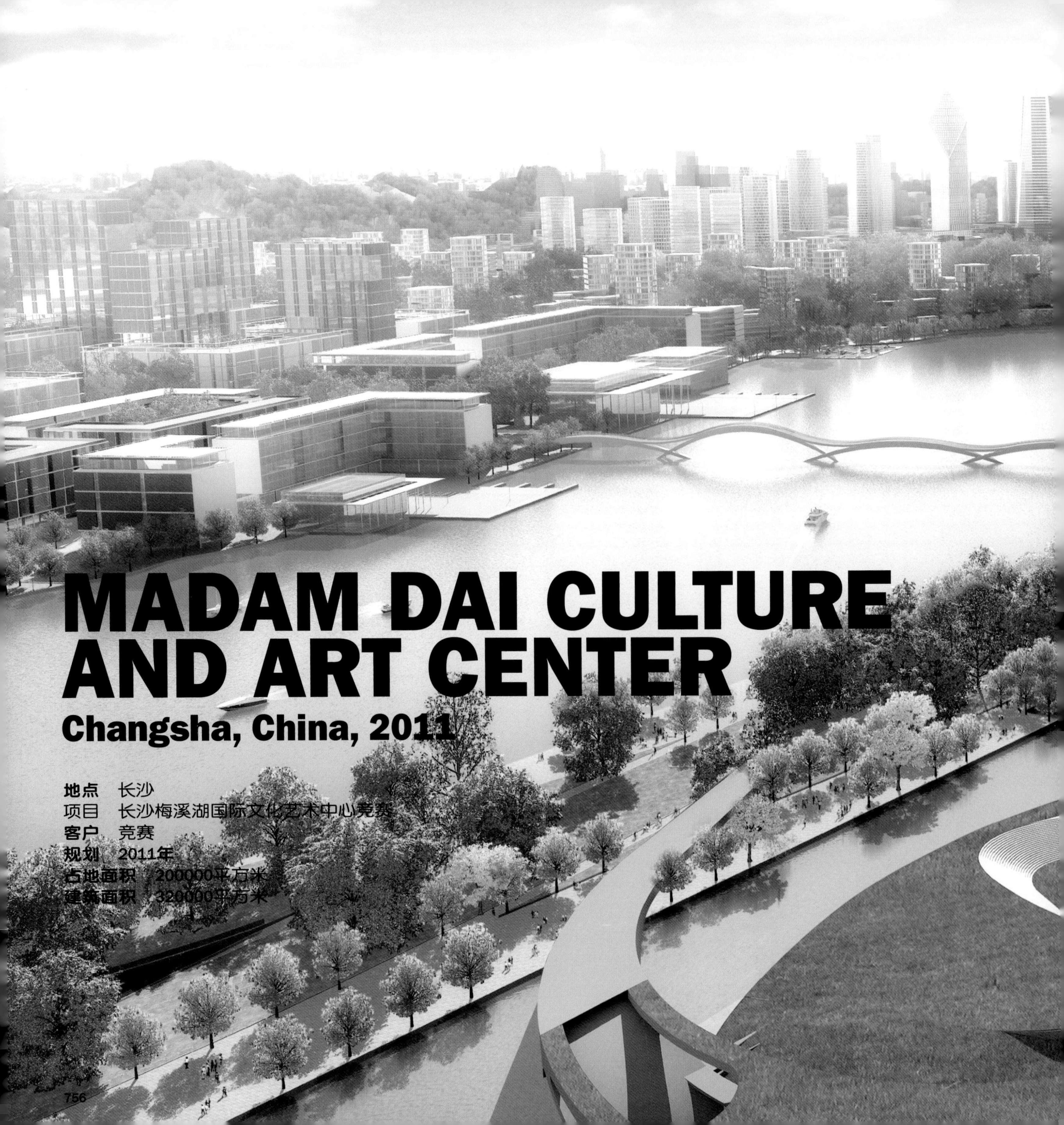

MADAM DAI CULTURE AND ART CENTER

Changsha, China, 2011

地点 长沙
项目 长沙梅溪湖国际文化艺术中心竞赛
客户 竞赛
规划 2011年
占地面积 200000平方米
建筑面积 320000平方米

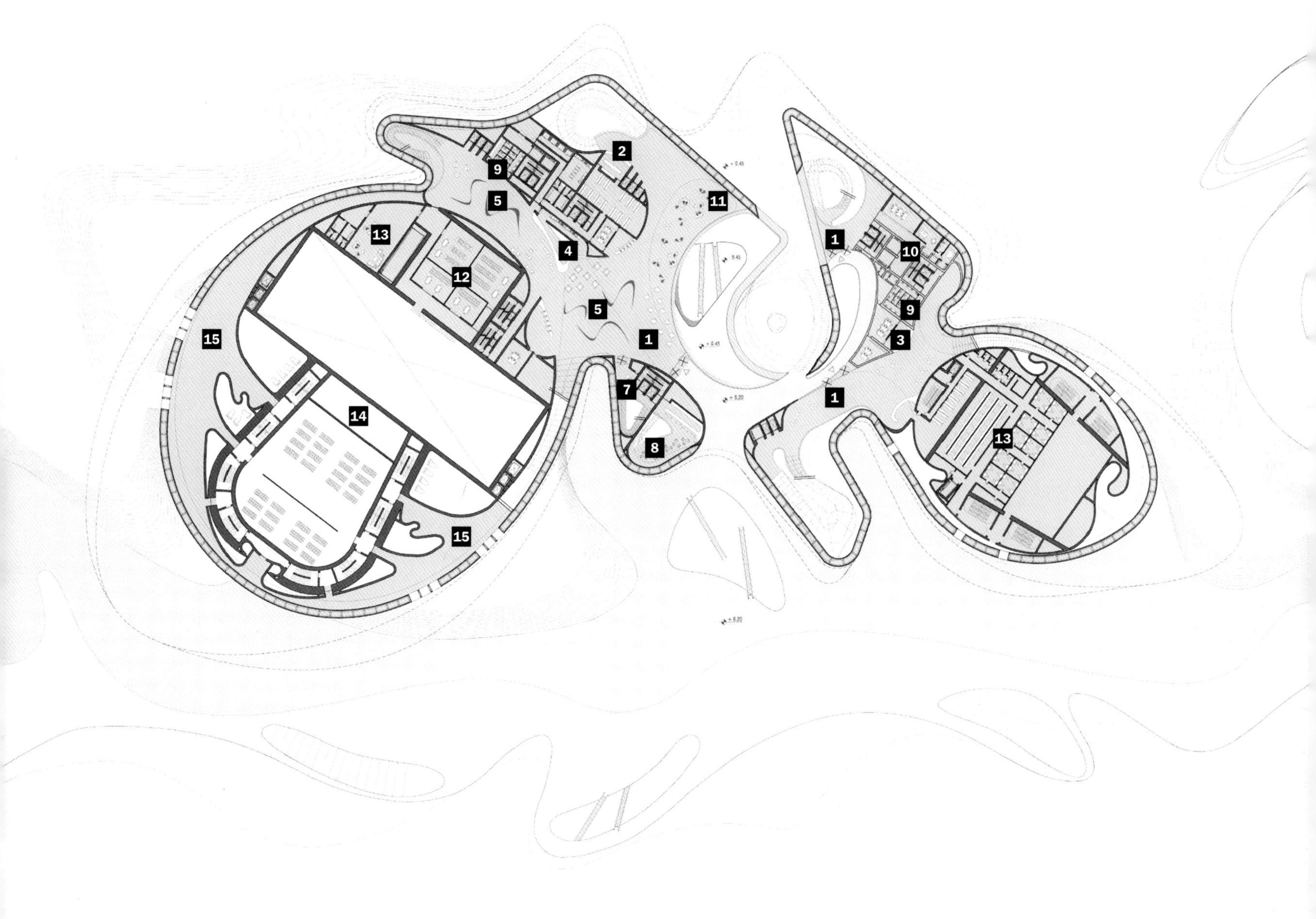

1. LOBBY / 前厅
2. CLOAKROOM / 衣帽间
3. TICKETS CENTER 售票中心
4. COFFEE HOUSE (TEA HOUSE) / 咖啡厅（茶室）
5. EXHIBITION OIL PAINTINGS / 油画展览
6. CONTROL TICKETS / 检票
7. VIP ENTRANCE / 贵宾入口
8. SHOP + COFFEE HOUSE / 商店+ 咖啡厅
9. BATHROOMS / 浴室
10. DORMITORY / 宿舍
11. LOUNGE / 休息厅
12. GENERAL STAGE MENAGMENT PART / 舞台总体管理
13. BACK STAGE OPERATION ROOMS / 后台业务用房
14. ORCHESTRA PIT / 乐池
15. SPACE FOR EVACUATE CROWD / 紧急疏散空间

0 10 20 m

ORIGINAL EMBRODERY DESIGN OF CHANGSHA
长沙原始图腾设计

MORPHING / 变形

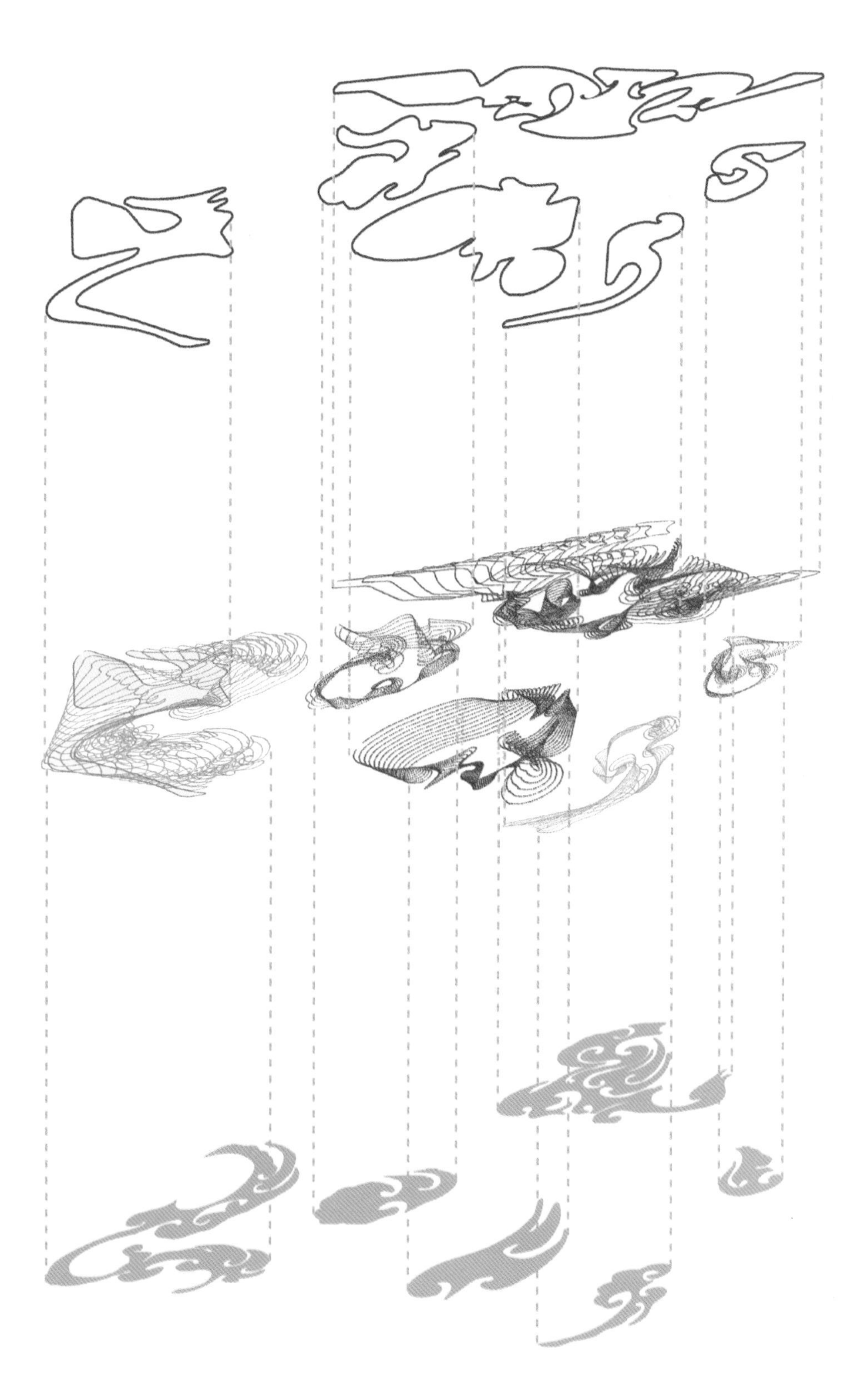

concept design 概念设计

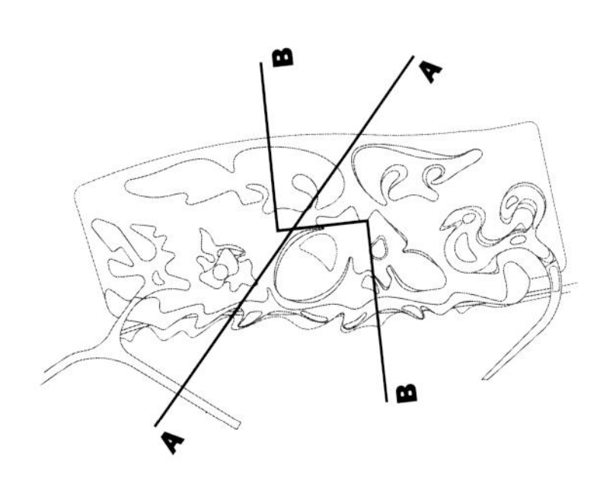

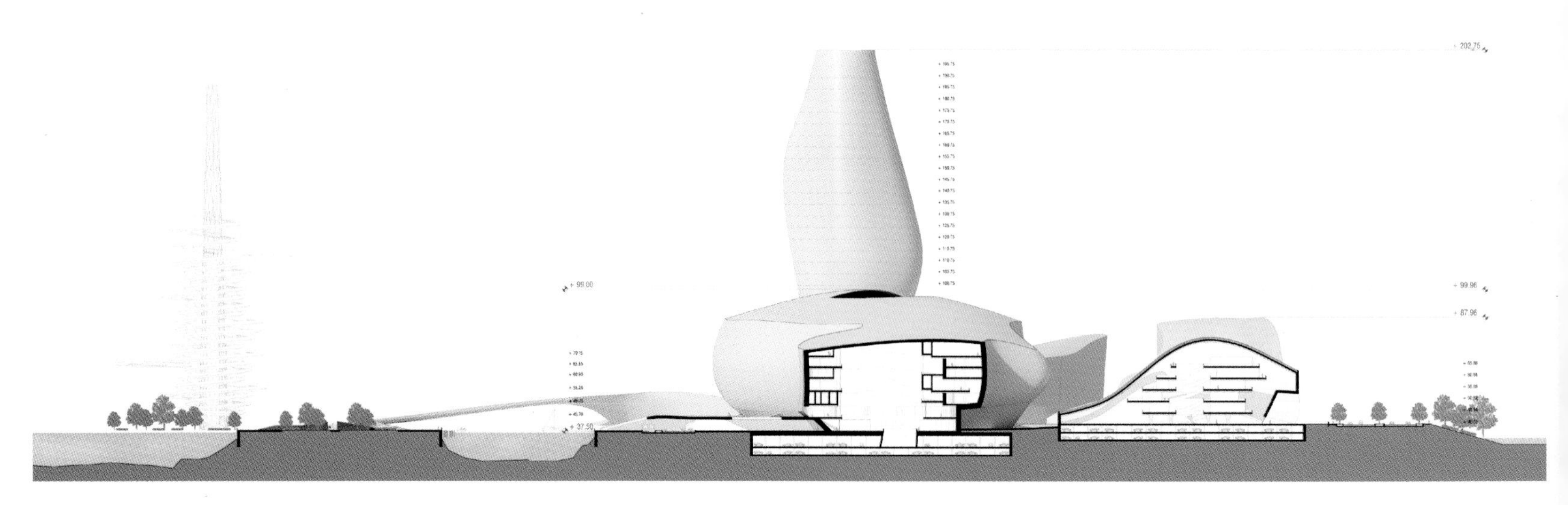

cross section A-A 横向剖面 A-A 1:100

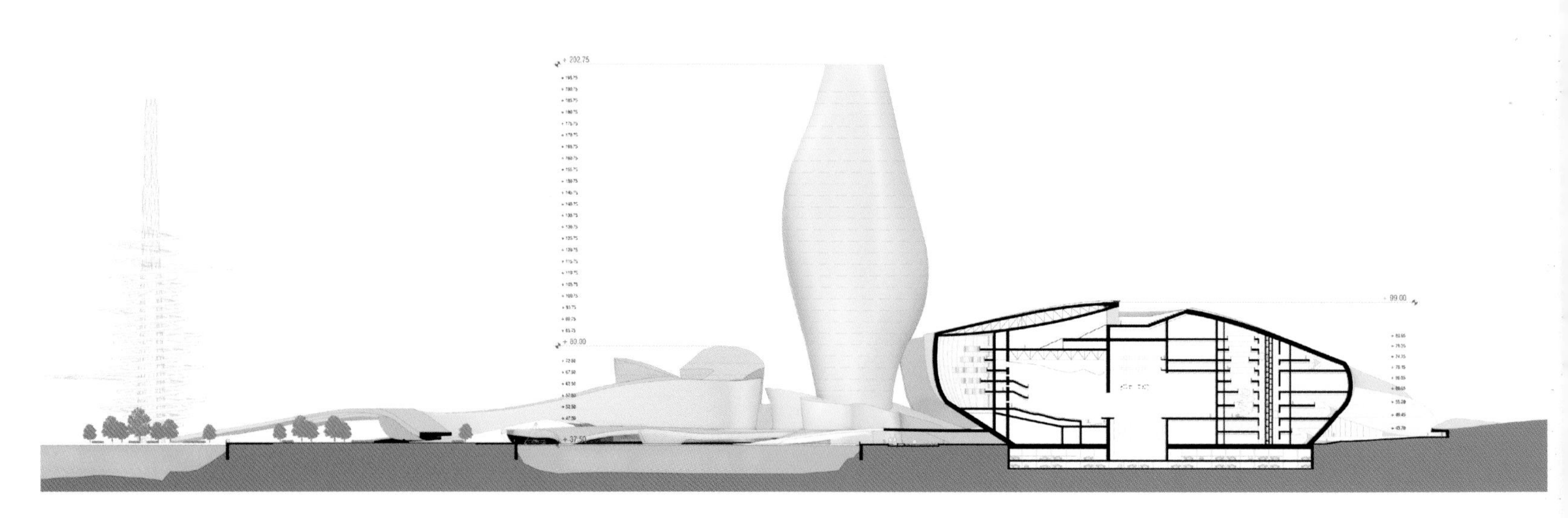

cross section B-B 横向剖面 B-B 1:100

0 25 50 m

INTERIORS

室内设计

BODY'S GYM

fitness center
Florence, Italy, 1998-2000

地点 佛罗伦萨，意大利
项目 健身中心
客户 Body's Gym S.r.l.
规划 1998年
建设 1999年-2000年
成本 700,000欧元
建筑面积 1,000平方米
体量 4,000立方米
承包商 Impresa Edile Turtora Cannizzaro

body's

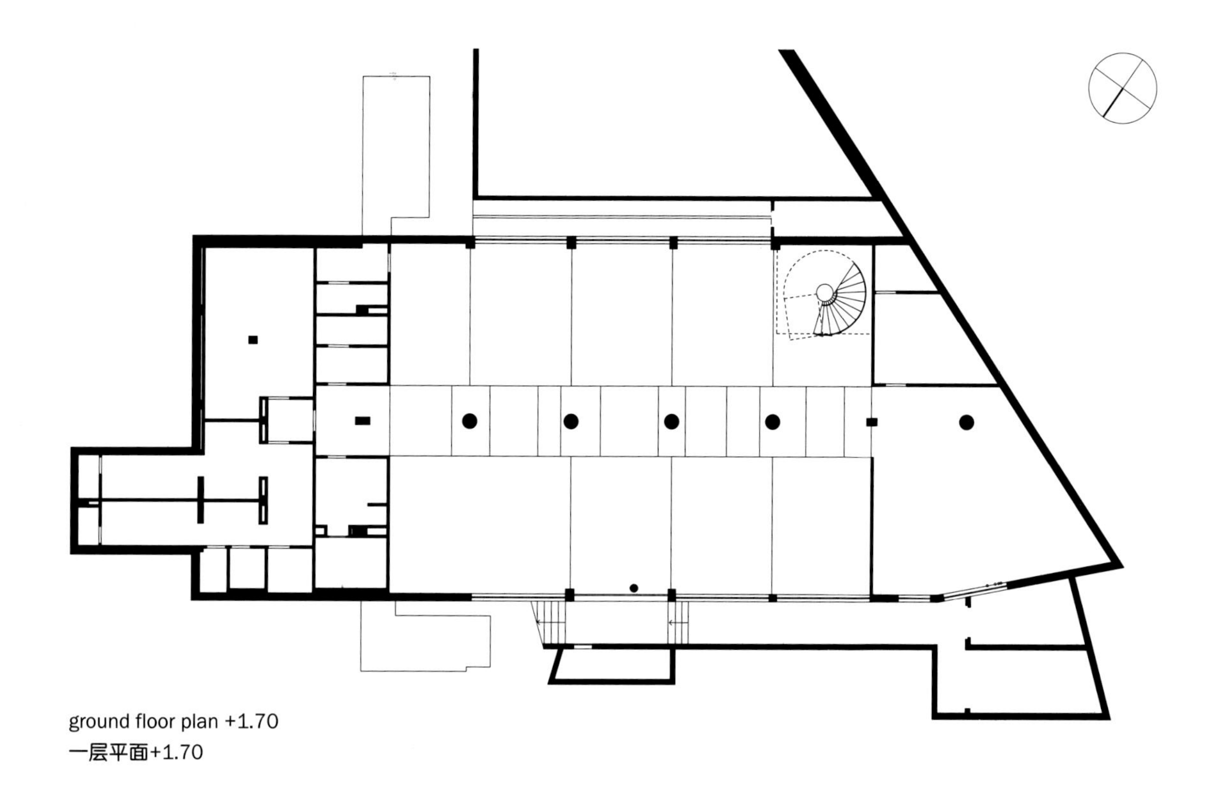

ground floor plan +1.70
一层平面+1.70

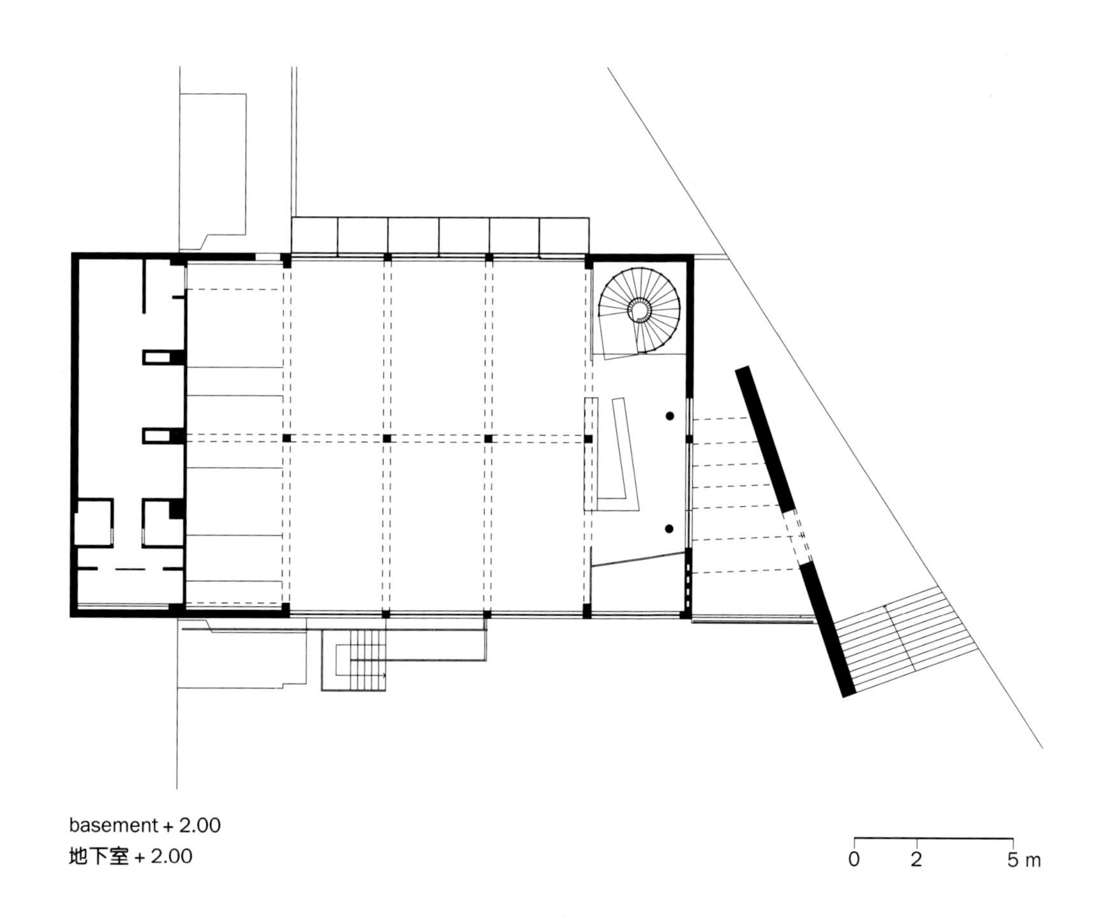

basement + 2.00
地下室 + 2.00

TORNABUONI ARTE GALLERY

Portofino, Genoa, Italy, 1999

地点 Portofino，热那亚，意大利
项目 商业，展览
客户 Tornabuoni Arte S.r.l.
规划 1999年
建设 1999年
成本 350,000欧元
建筑面积 30平方米
承包商 Fratelli Giani S.r.l.

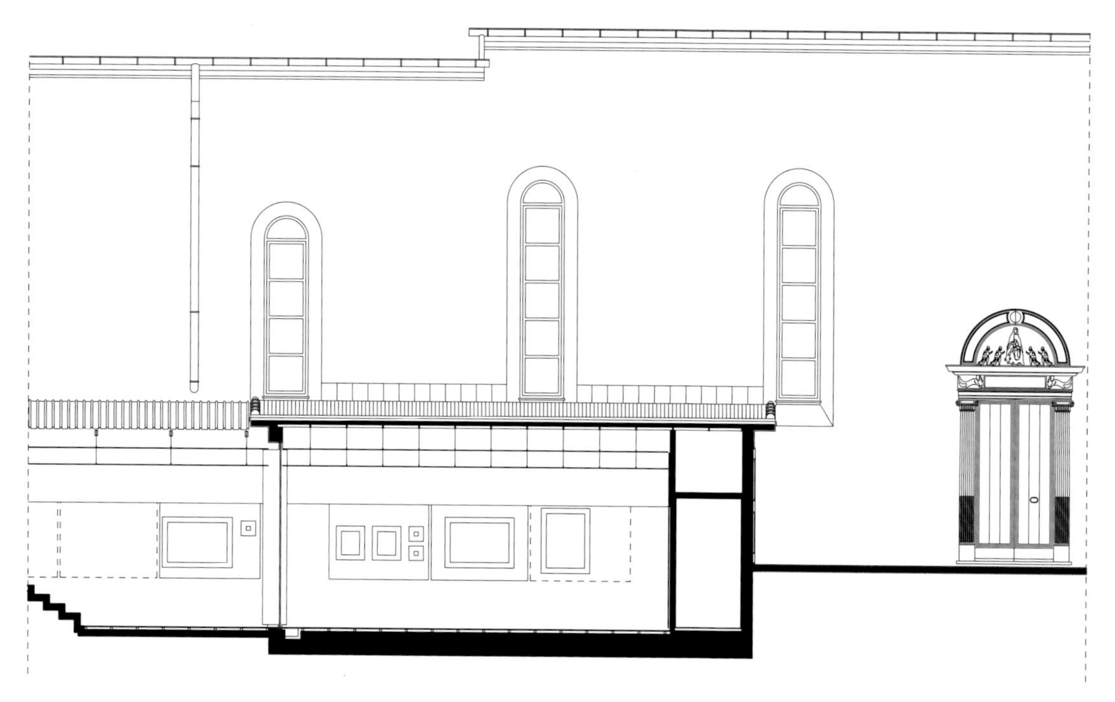

section A-A'
A-A'剖面

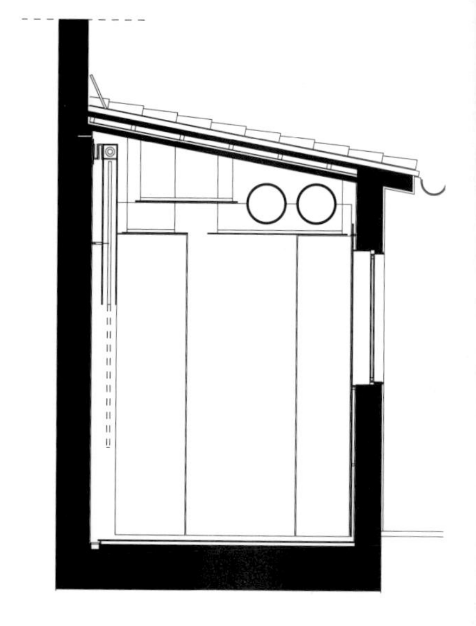

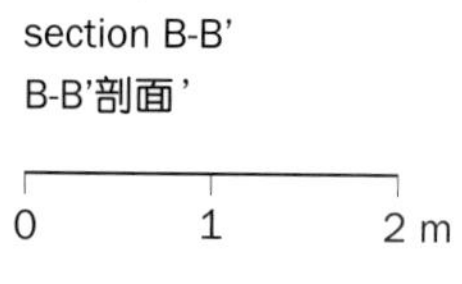

section B-B'
B-B'剖面'

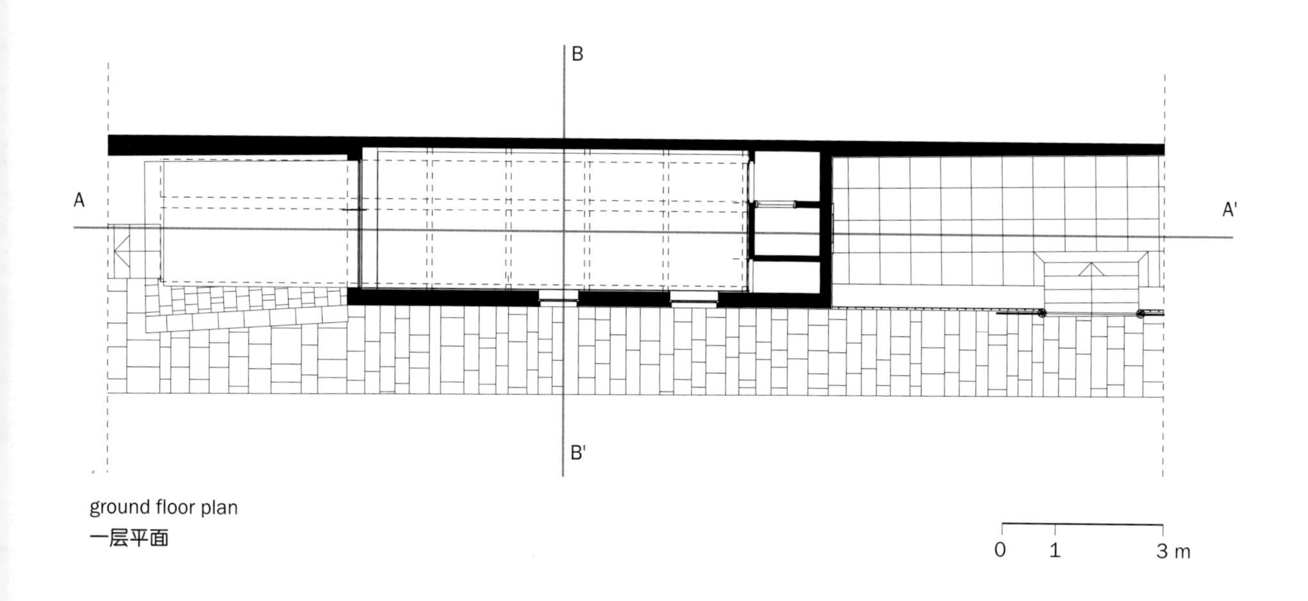

ground floor plan
一层平面

GRANITIFIANDRE SHOWROOM

Castellarano, Reggio Emilia, Italy, 2001

地点　Castellarano，Reggio Emilia，意大利
项目　陈列室
客户　GranitiFiandre S.p.A.
结构　Studio Tecnico Cuoghi
规划　2001年
建设　2001年

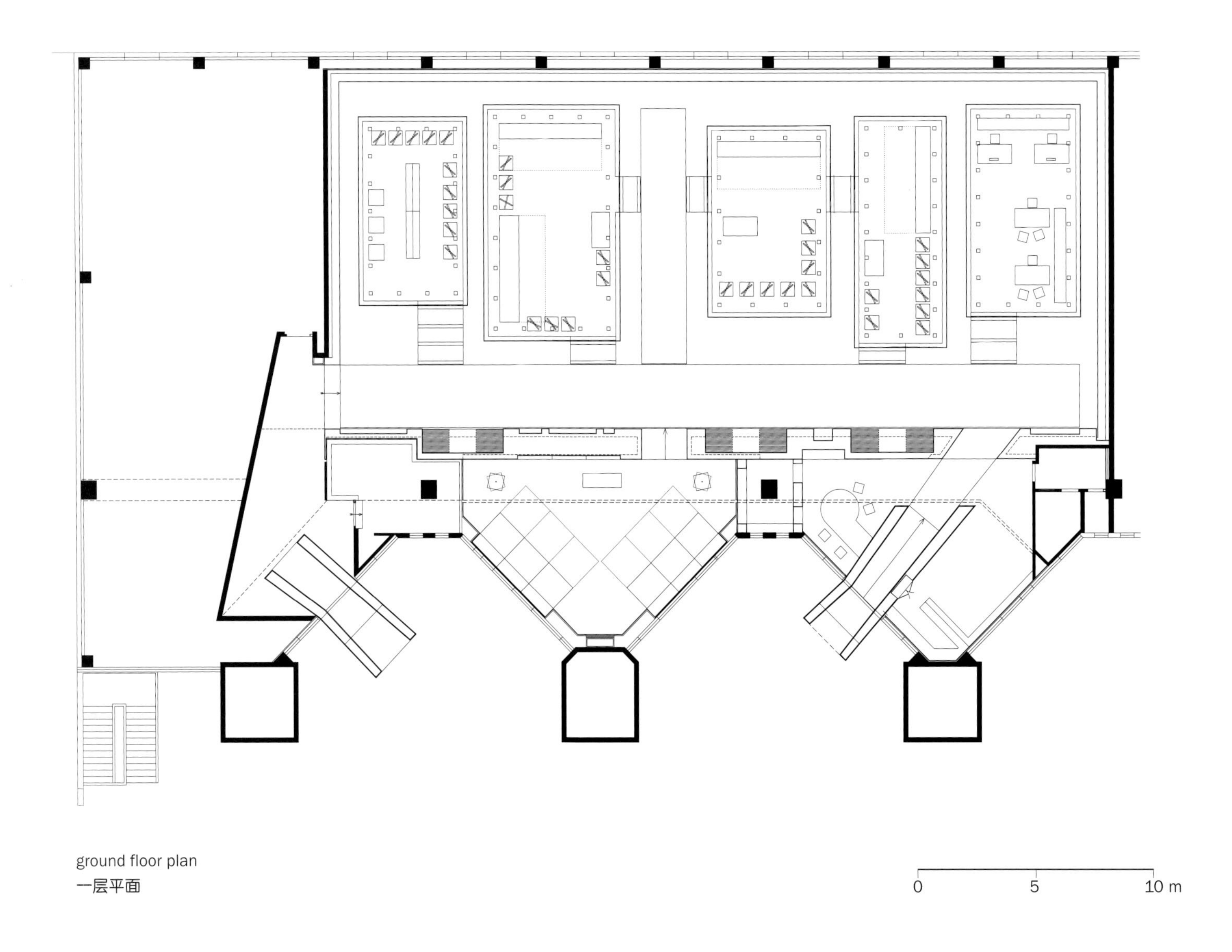

ground floor plan

一层平面

BALUARDO SAN COLOMBANO

restaurant and lounge bar
Lucca, Italy, 2002-2003

地点　卢卡，意大利
项目　餐馆和雅座酒吧
客户　Carmafrigor S.r.l.
系统　P.I. Luca Pollastrini
规划　2002年-2003年
建设　2003年
成本　600,000欧元
建筑面积　200平方米
承包商　Michele Bianchi S.r.l.

AN
CO
LOM
BANO

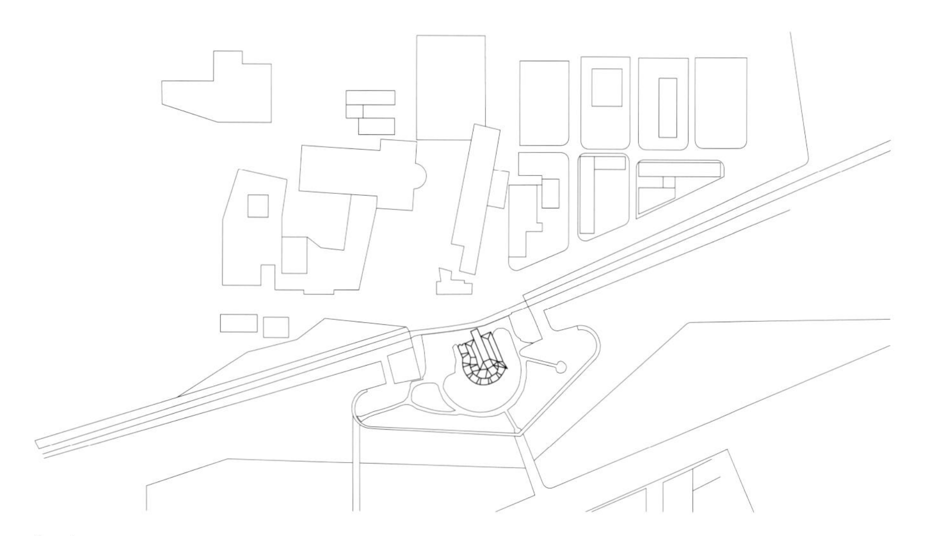

site plan
位置图

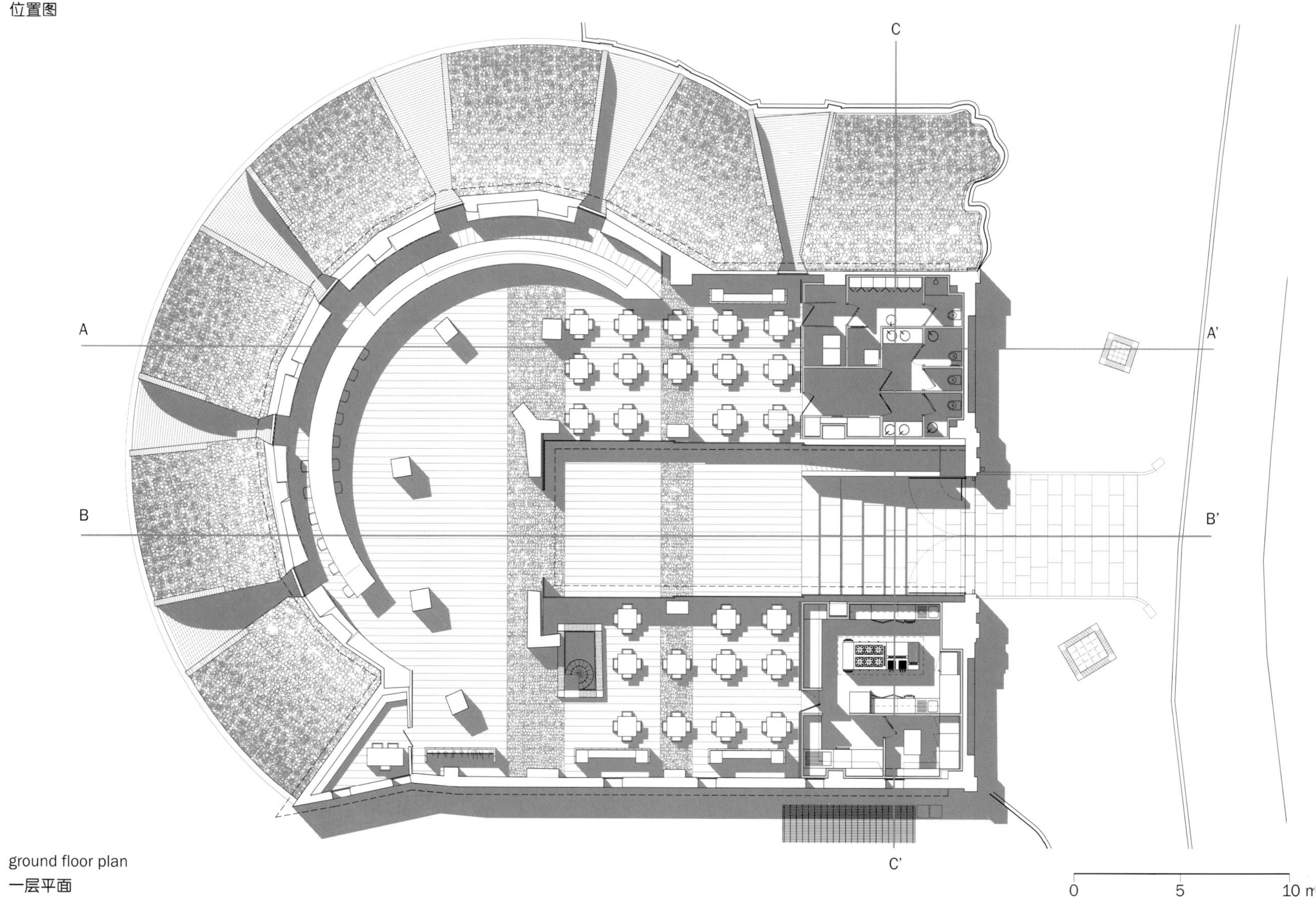

ground floor plan
一层平面

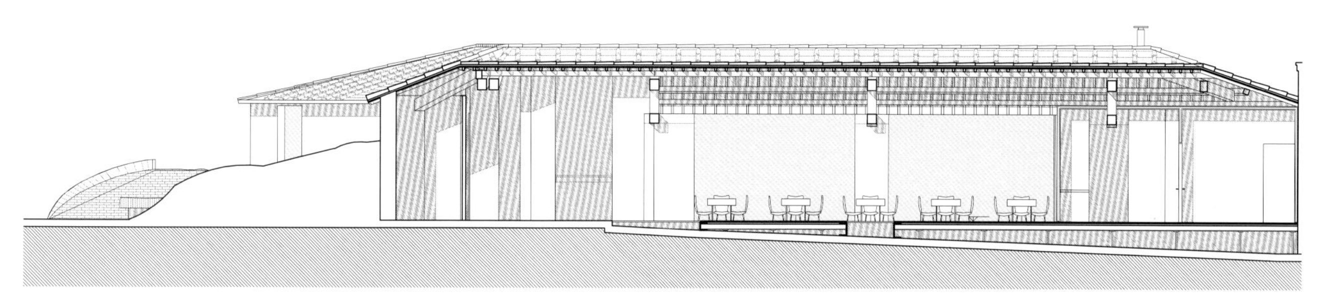

section A-A'
A-A'剖面

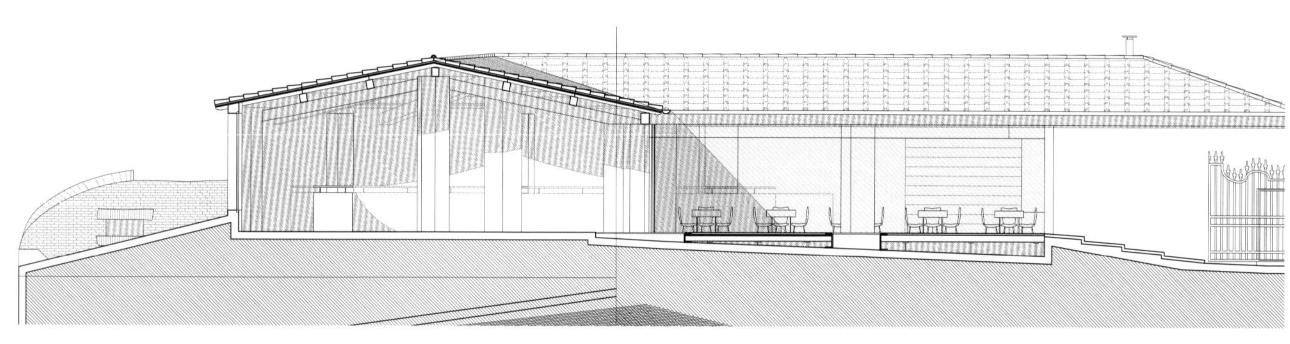

section B-B'
B-B'剖面

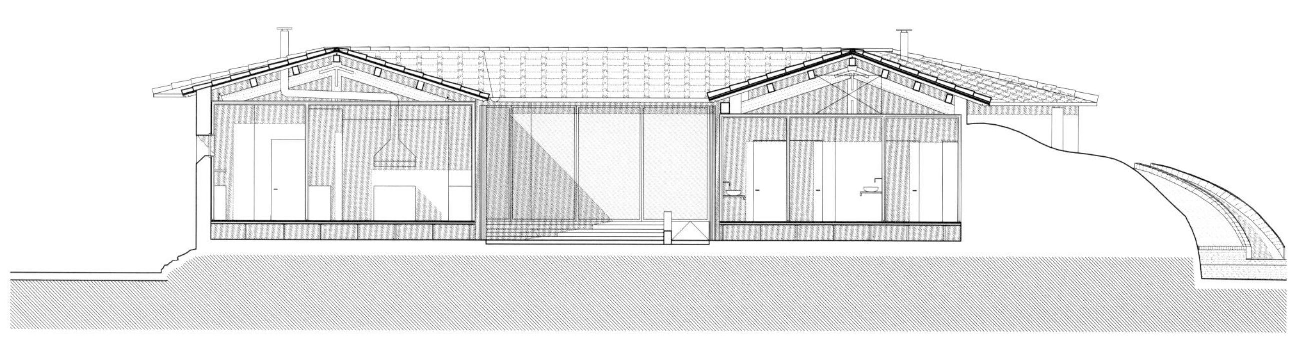

section C-C'
C-C'剖面

MATEUS

TORNABUONI ARTE GALLERY

Venice, Italy, 2004

地点 威尼斯，意大利
项目 商业，展览
客户 Tornabuoni Arte S.r.l.
结构 Favero&Milan Ingegneria
规划 2004年
建设 2004年
成本 700,000欧元
建筑面积 60平方米
承包商 Friulan

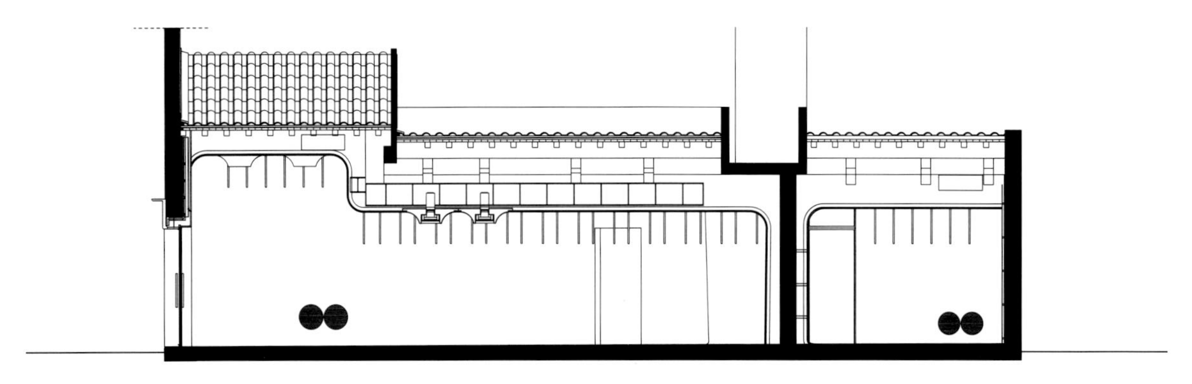

section A-A'
A-A'剖面

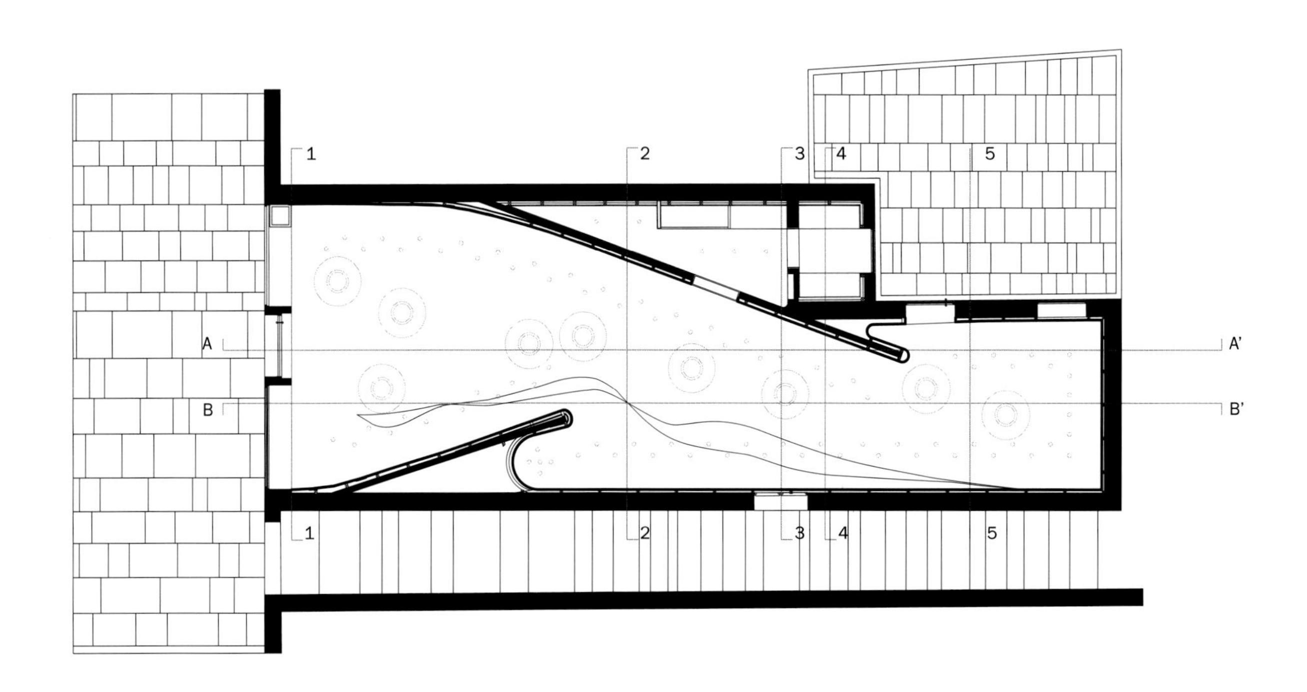

ground floor plan
一层平面

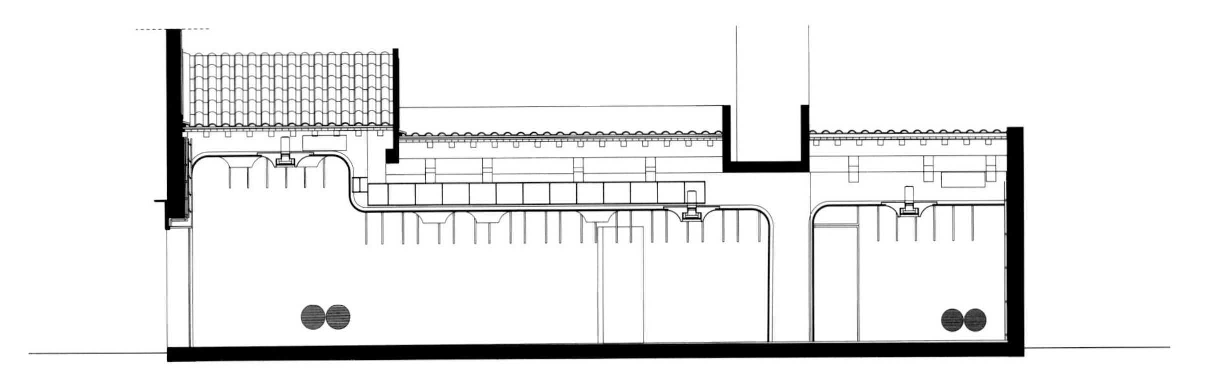

section B-B'
B-B'剖面

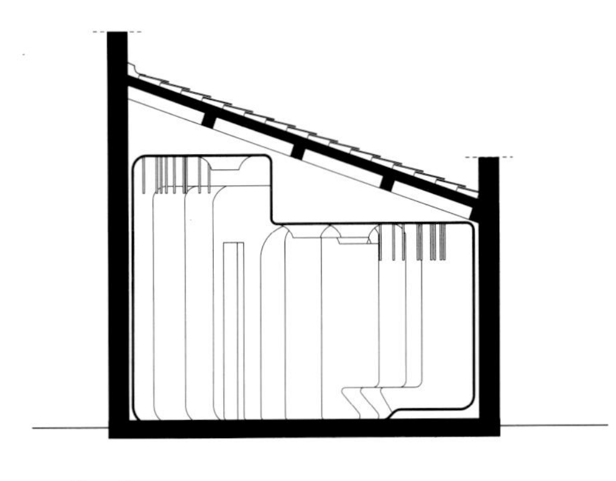

section 1
1剖面

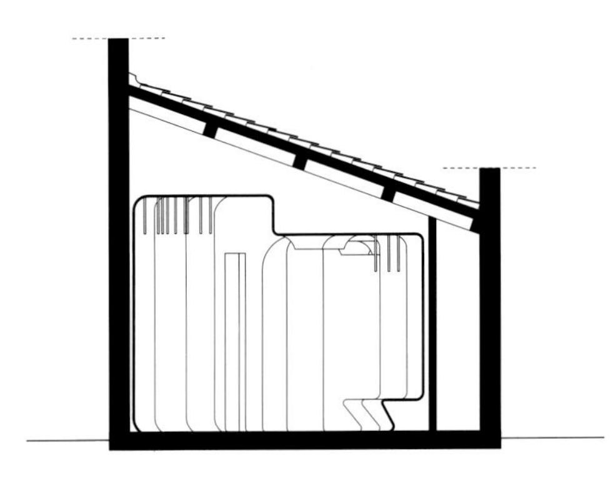

section 2
2剖面

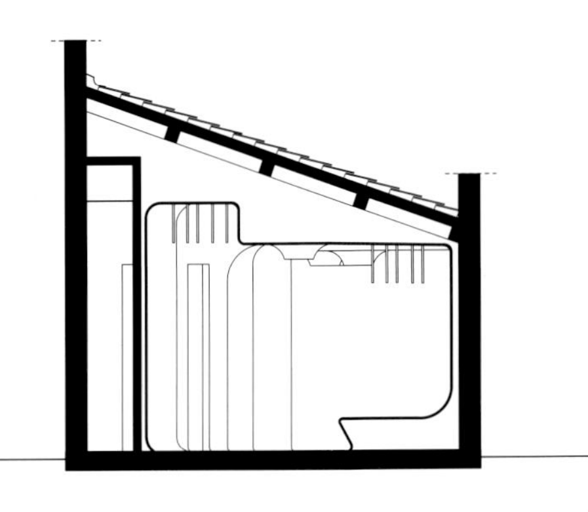

section 3
3剖面

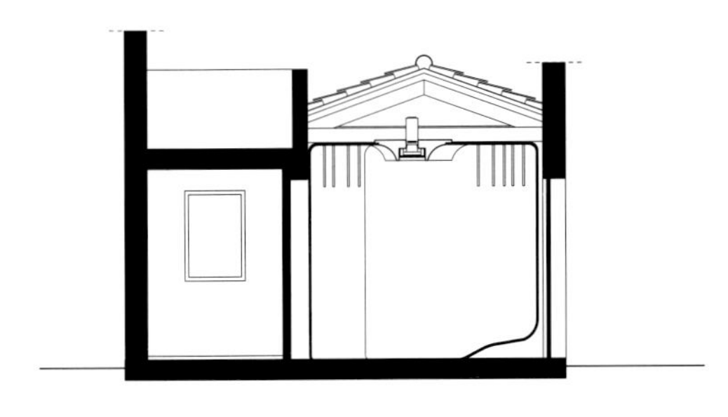

section 4
4剖面

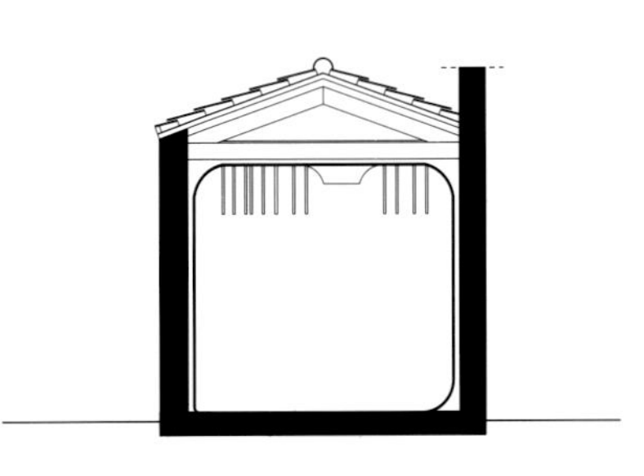

section 5
5剖面

SETS

装置

LONELY LIVING

Venice Biennial, Venice, Italy, 2002

地点 Giardini di Castello，威尼斯，意大利
项目 展览，装置
客户 Aid'Associazione italiana di Architettura
结构 aei progetti：Niccolò De Robertis
系统 Mart[illegible] Mantovani
规划 2002年
建设 2002年
建筑面积 900平方米
体量 270立方米
承包商 Fima Cosma Silos

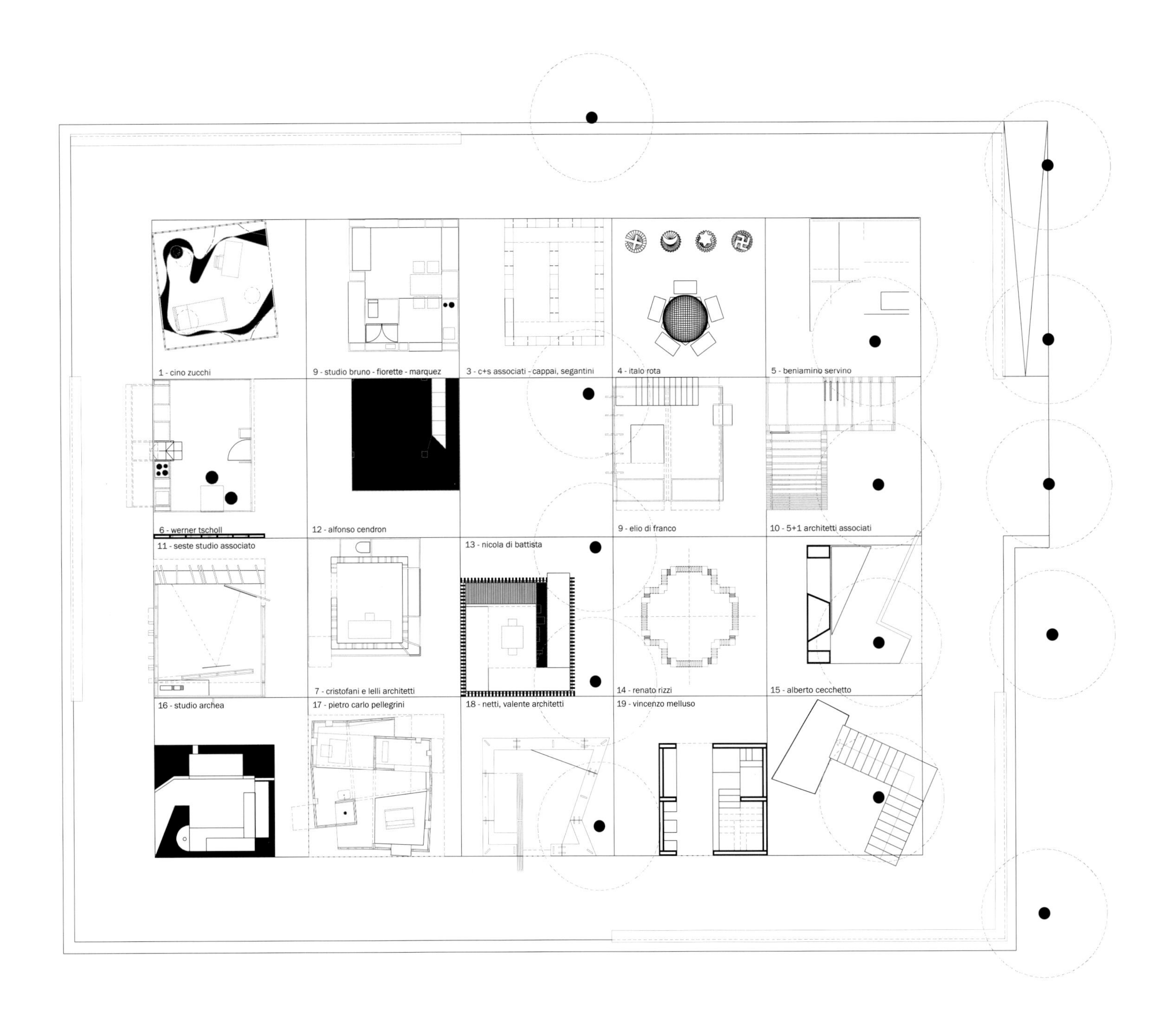

site plan
位置图

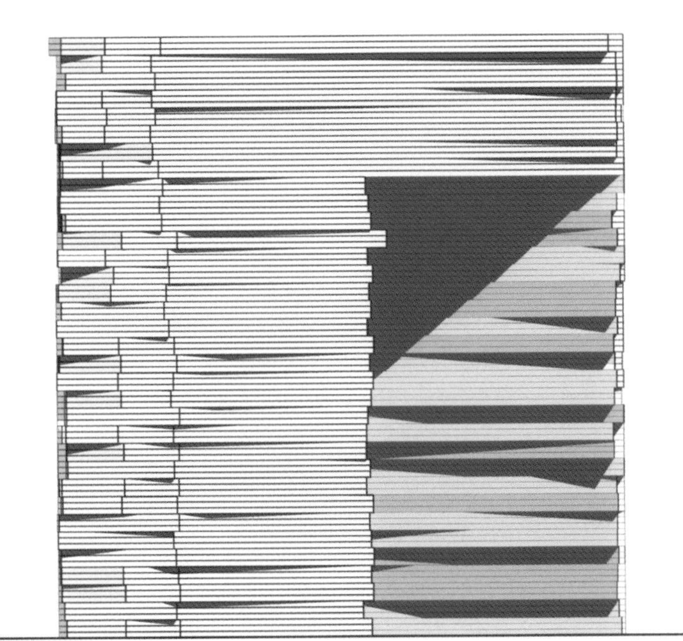

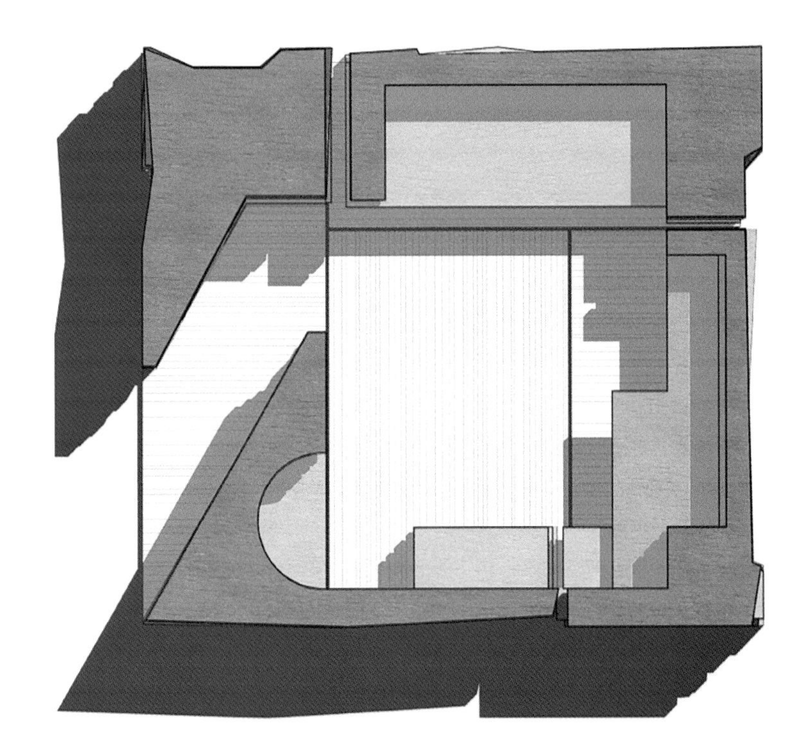

elevations and plan
立面和平面

ENZIMI

Rome, Italy, 2003

地点 罗马，意大利
项目 展览，装置
客户 Zone Attive – Comune di Roma
规划 2003年
建设 2003年
成本 15,000欧元
铝制品面积 1,600平方米
覆盖聚碳酸酯的面积 1,884平方米
体量 2,240立方米
承包商 Nolostand

INFOPOINT
ENZIMI 2003 ROMA
atac
es.hotel
ExibArt
ALIAS
COLORS
TISCALI
ELISEO
CERTE SIGNORE FAREBBERO
QUALSIASI COSA PER UNA
BUONA CAUSA
calendar
girls

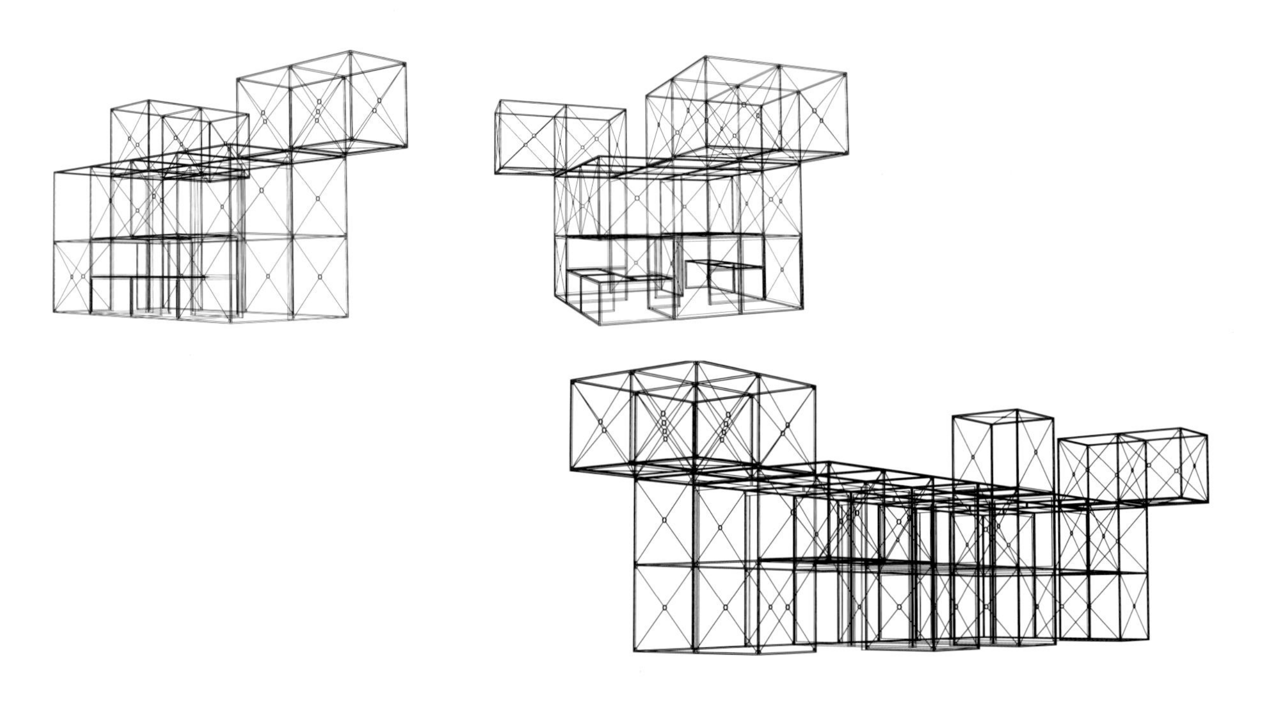

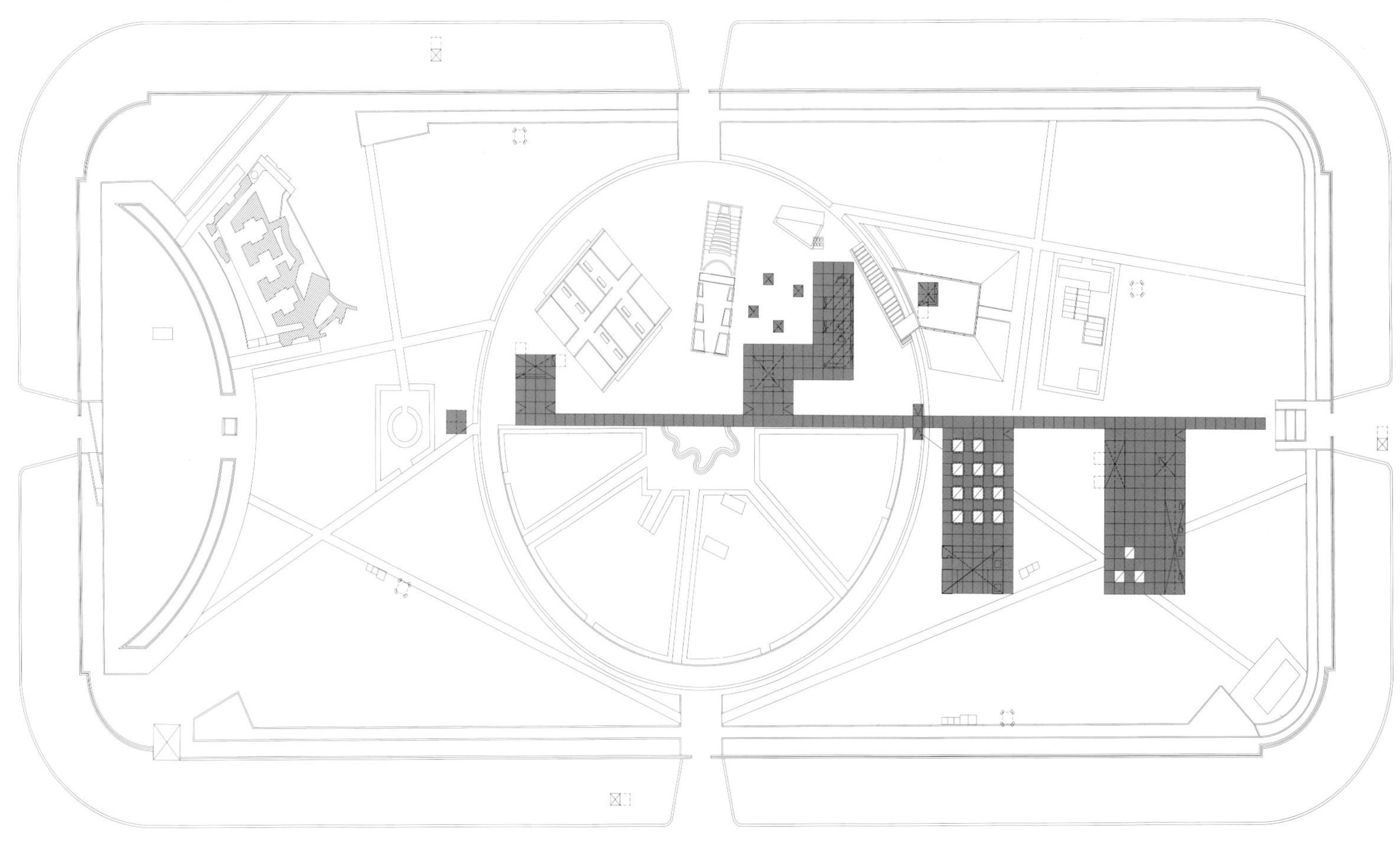

site plan
位置图

EN
ZI
MI
20
03
RO

EASTPAK

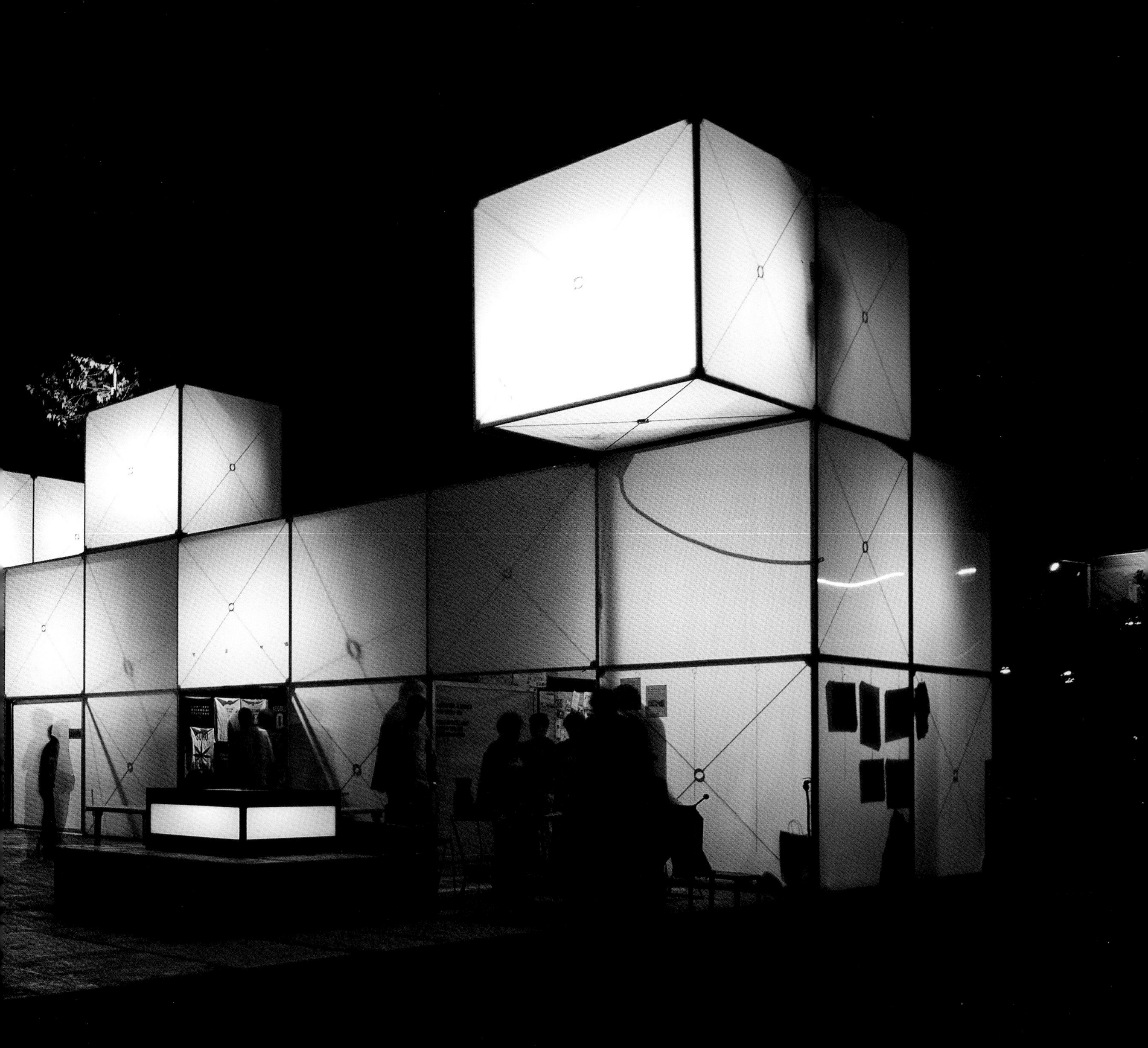

VINAR

Stazione Leopolda, Florence, Italy, 2005

地点 Stazione Leopolda，佛罗伦萨，意大利
项目 装置
客户 Federico Motta Editore and Pitti Immagine
规划 2005年
建设 2005年
成本 300,000欧元
建筑面积 600平方米
承包商 Machina S.r.l.

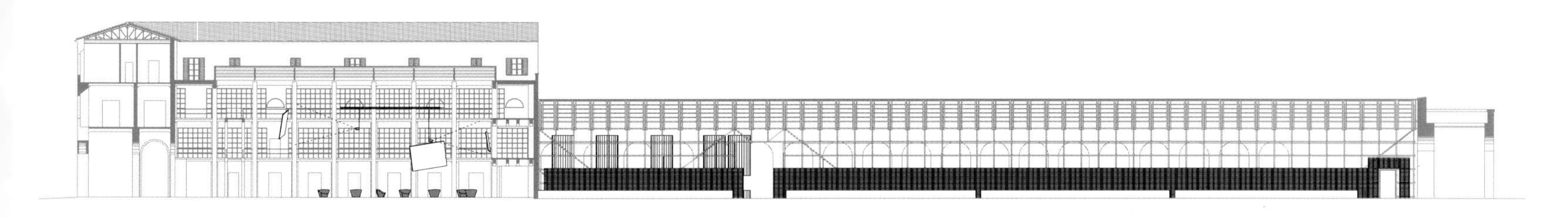

section A-A'
A-A'剖面

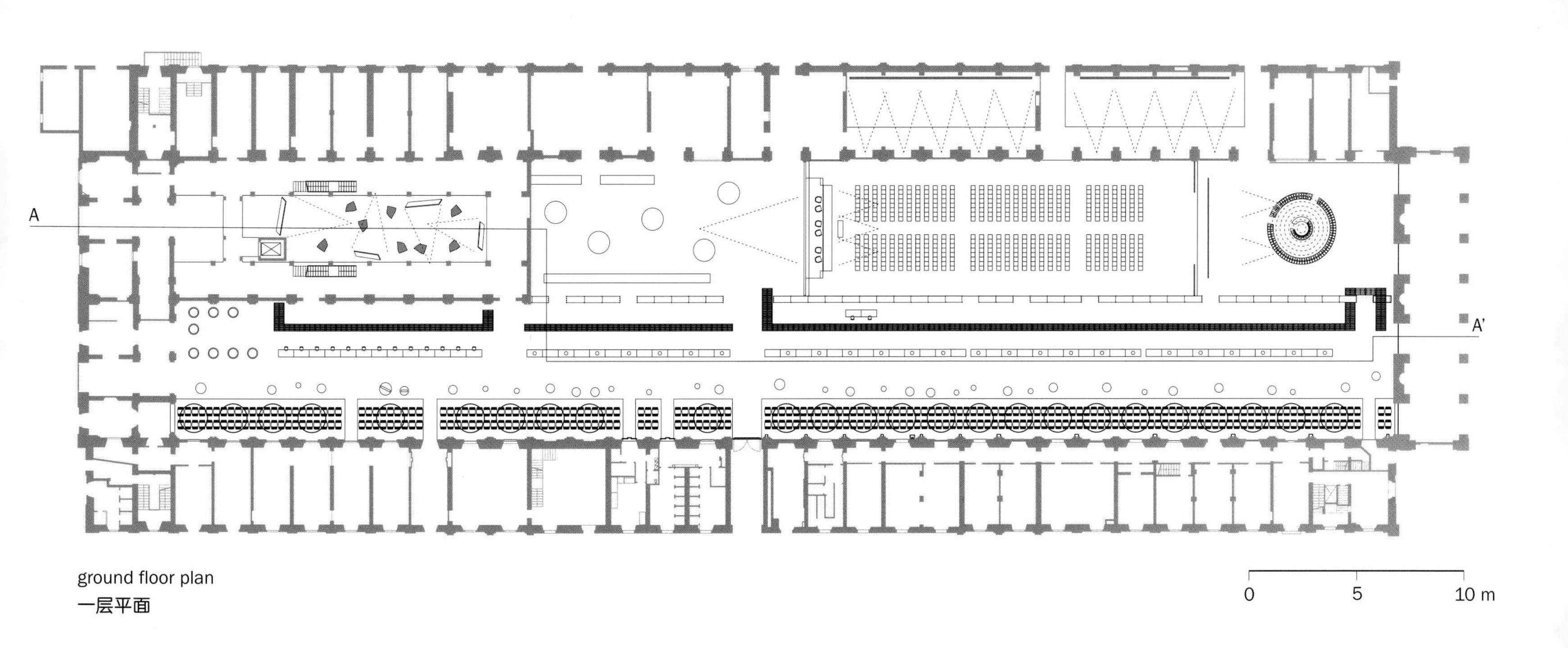

ground floor plan
一层平面

Westbank, Canada 2000
CANTINE
architetture

OSSI Cadenazzo, Ticino, Svizzera / 1994

LABORATORIO ITALIA

Rome, Italy, 2006

地点 罗马，意大利
项目 展览，装置
客户 Aid'A Agenzia Italiana d'Architettura, DARC Direzione Generale per l'Architettura e l'Arte Contemporanee
规划 2006年
建设 2006年

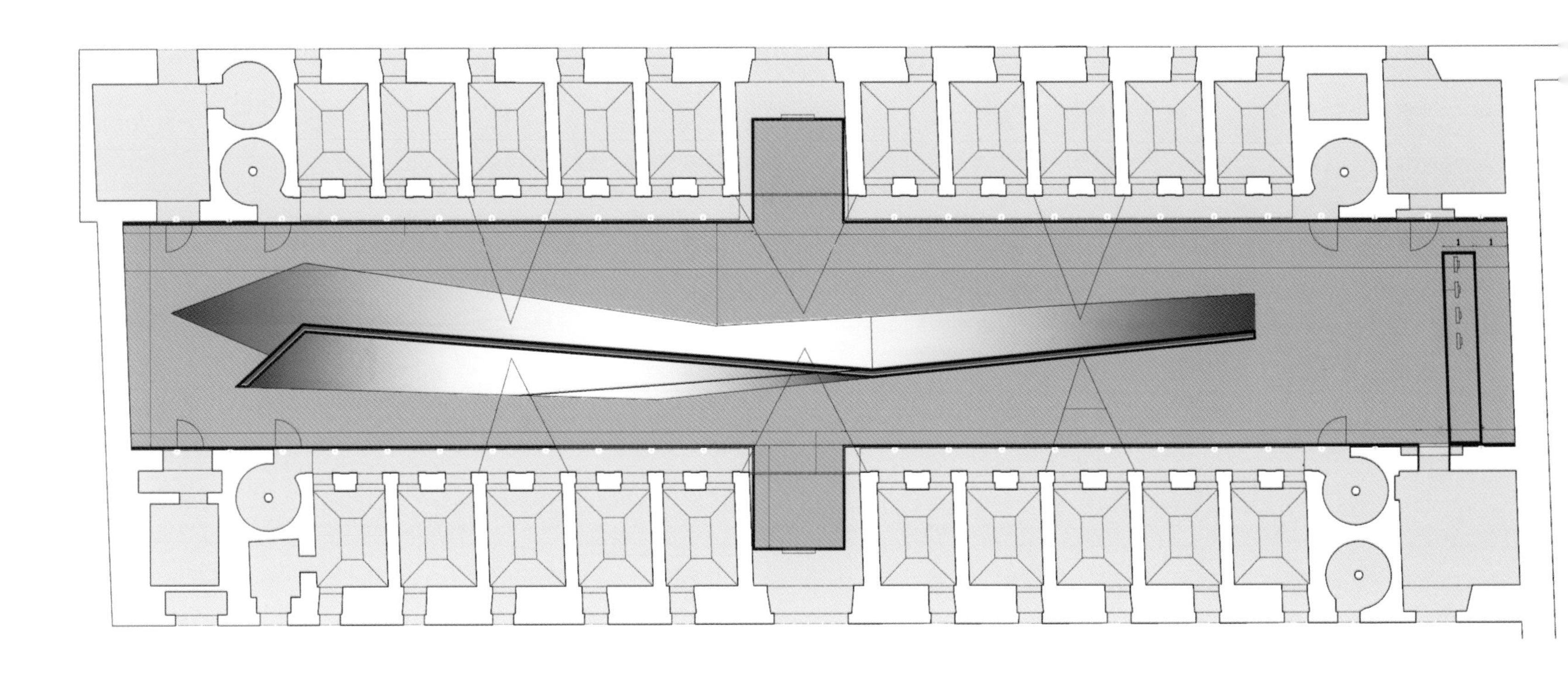

general plan
总体平面

0 1 5 m

ANNALI DELLA ARCHITETTURA 2007

Palazzo Reale, Naples, Italy, 2007

地点 Palazzo Reale，那不勒斯，意大利
项目 展览，装置
客户 Fondazione Annali dell'Architettura e delle Città, Napoli
规划 2007年
建设 2007年
承包商 Domino S.r.l.

06
07
08
msc
MEDITERRANEAN SHIPPING CO

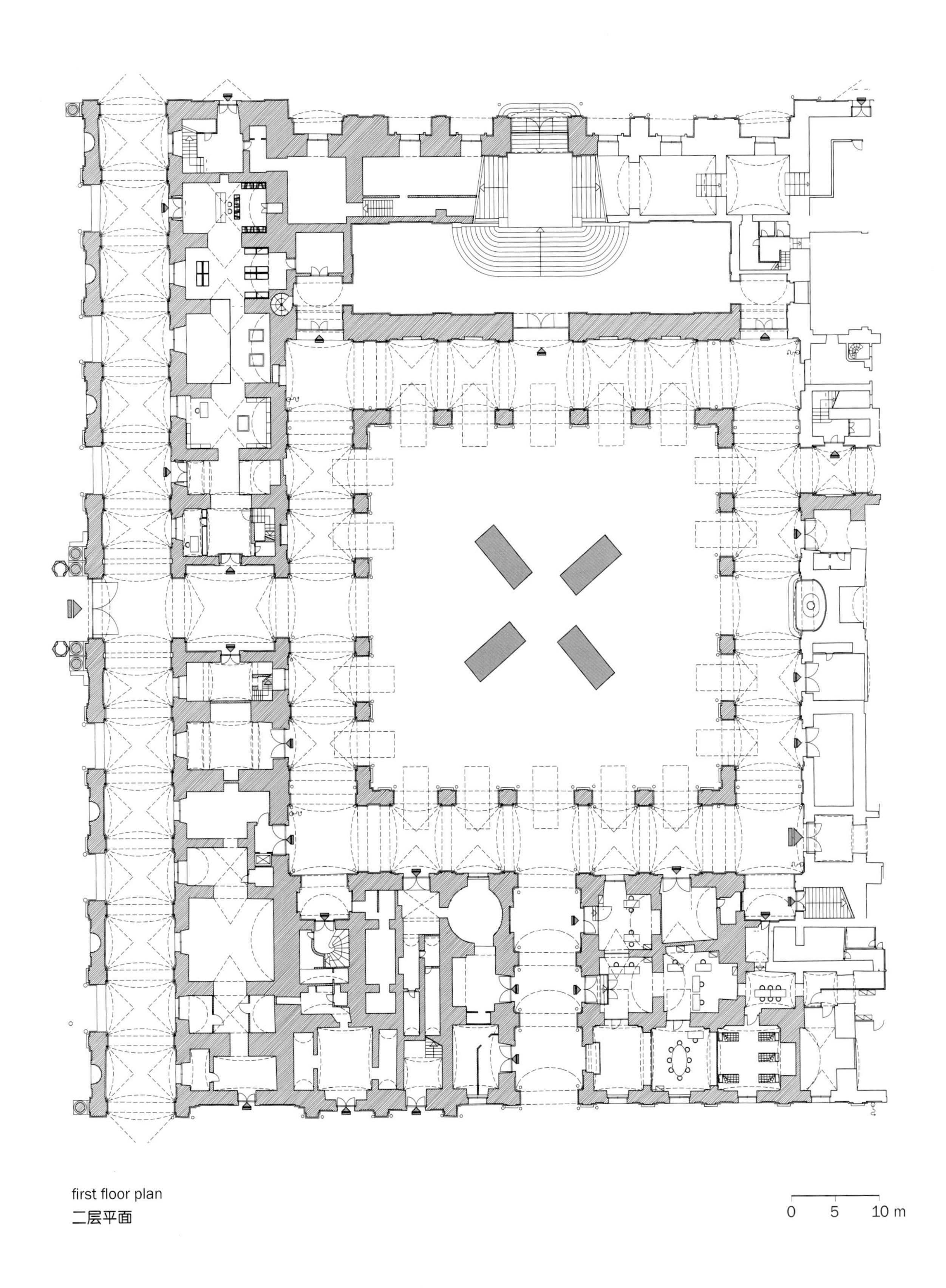

first floor plan
二层平面

msc
MSCU
602953 8
22G1
M. G. W.
TARE
NET
CU.CAP.
B
MEDU
msc
M. G. W.
TARE
NET
CU.CAP.

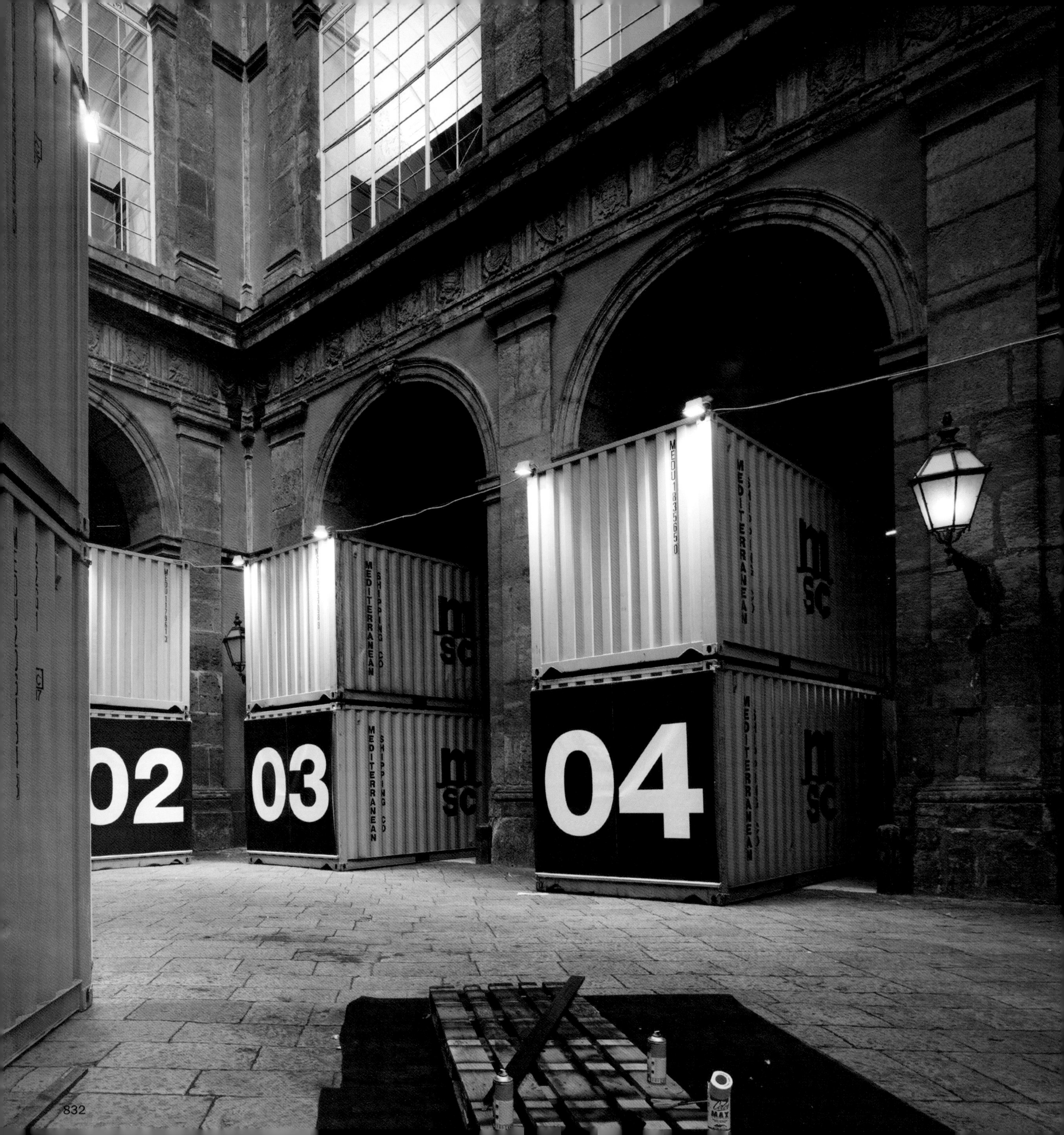
02
03
04
MEDITERRANEAN
SHIPPING CO
msc

MEDITERRANEAN SHIPPING CO
MEDITERRANEAN SHIPPING CO
A
07
MSCU
M.G.W
TARE
NET
CAP.

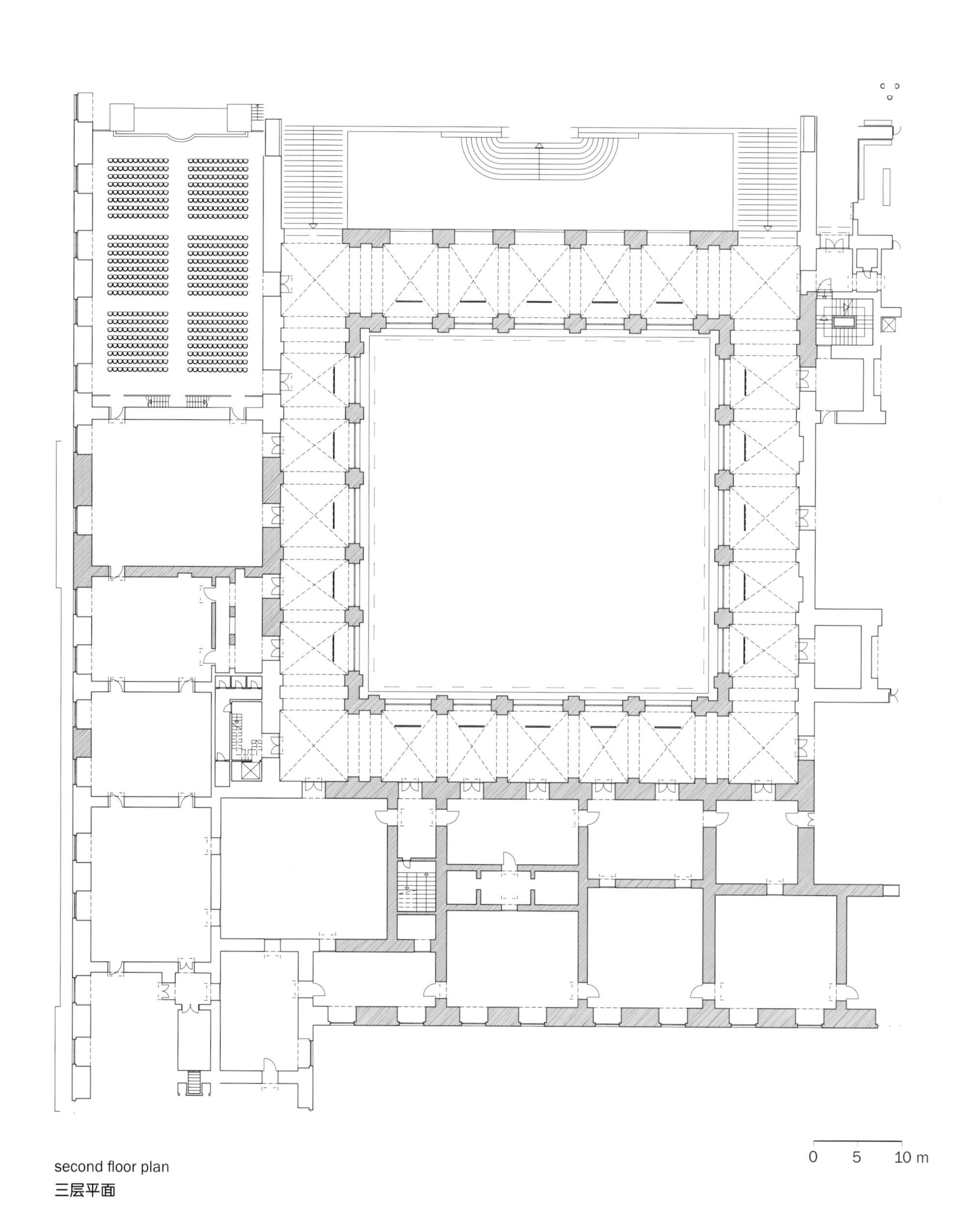

second floor plan
三层平面

NIA / NEPTUNIA 1950
NEPTUNIA 1932 / OCEANIA 1933

1968/2008 QUARANT'ANNI DI DESIGN

Milan, Italy, 2008

地点 Il Sole 24 ORE总部，米兰，意大利
项目 展览，装置
客户 Il Sole 24 ORE Business Media
规划 2008年
建设 2008年
承包商 Machina S.r.l.

1968 2008
quaranta anni di design
tra continuità e discontin

dataton
Display 3

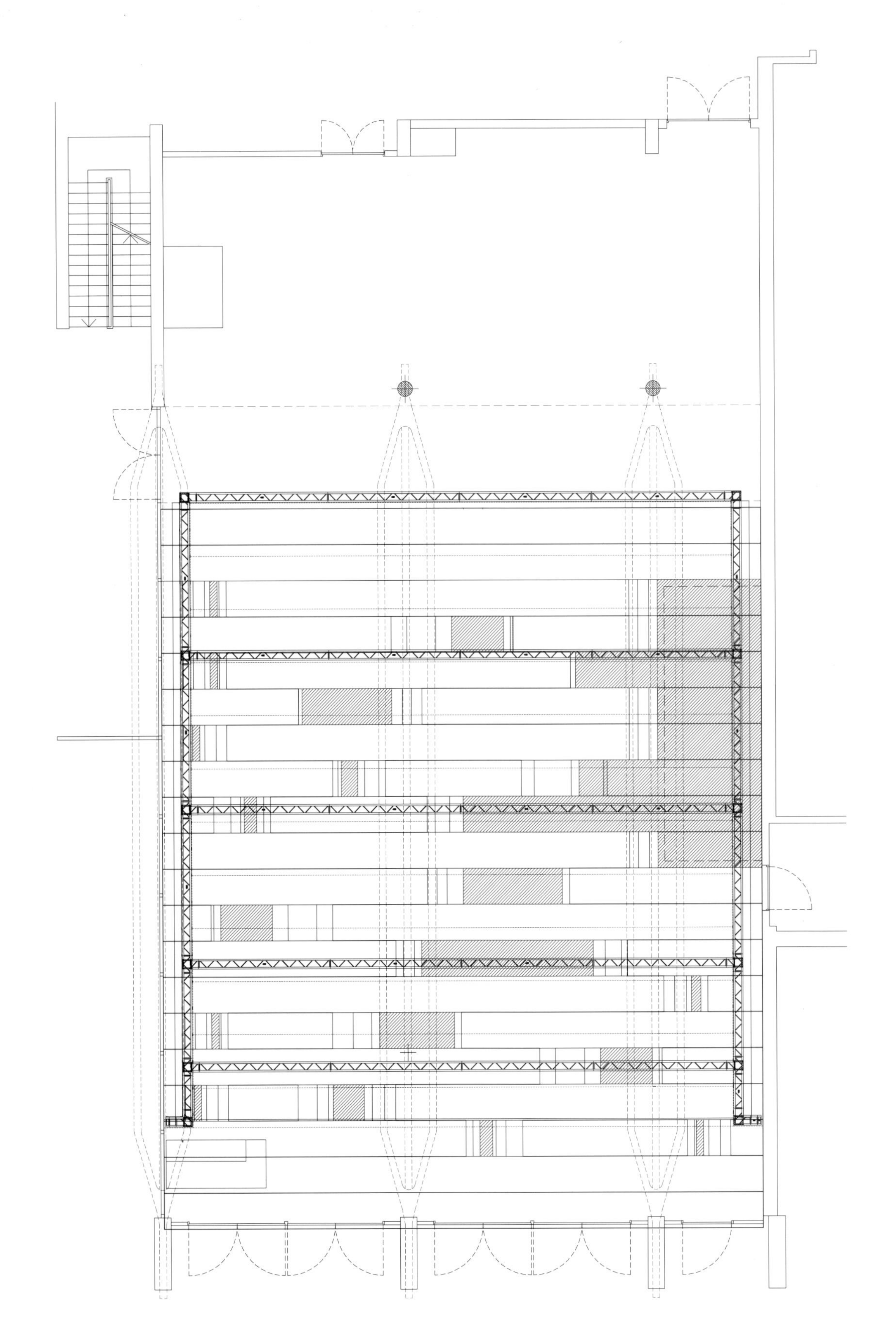

plan
平面

Branzi

USCITA DI EMERGENZA
USCITA DI EMERGENZA

ANNALI DELLA ARCHITETTURA 2008

Palazzo Reale, Naples, Italy, 2008

地点　Palazzo Reale，那不勒斯，意大利
项目　展览，装置
客户　Fondazione Annali dell'Architettura e delle Città, Napoli
规划　2008年
建设　2008年

emergenza

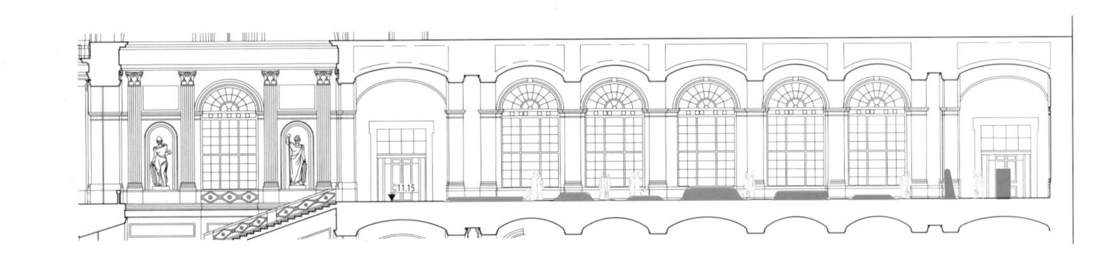

sezione longitudinale
纵剖面

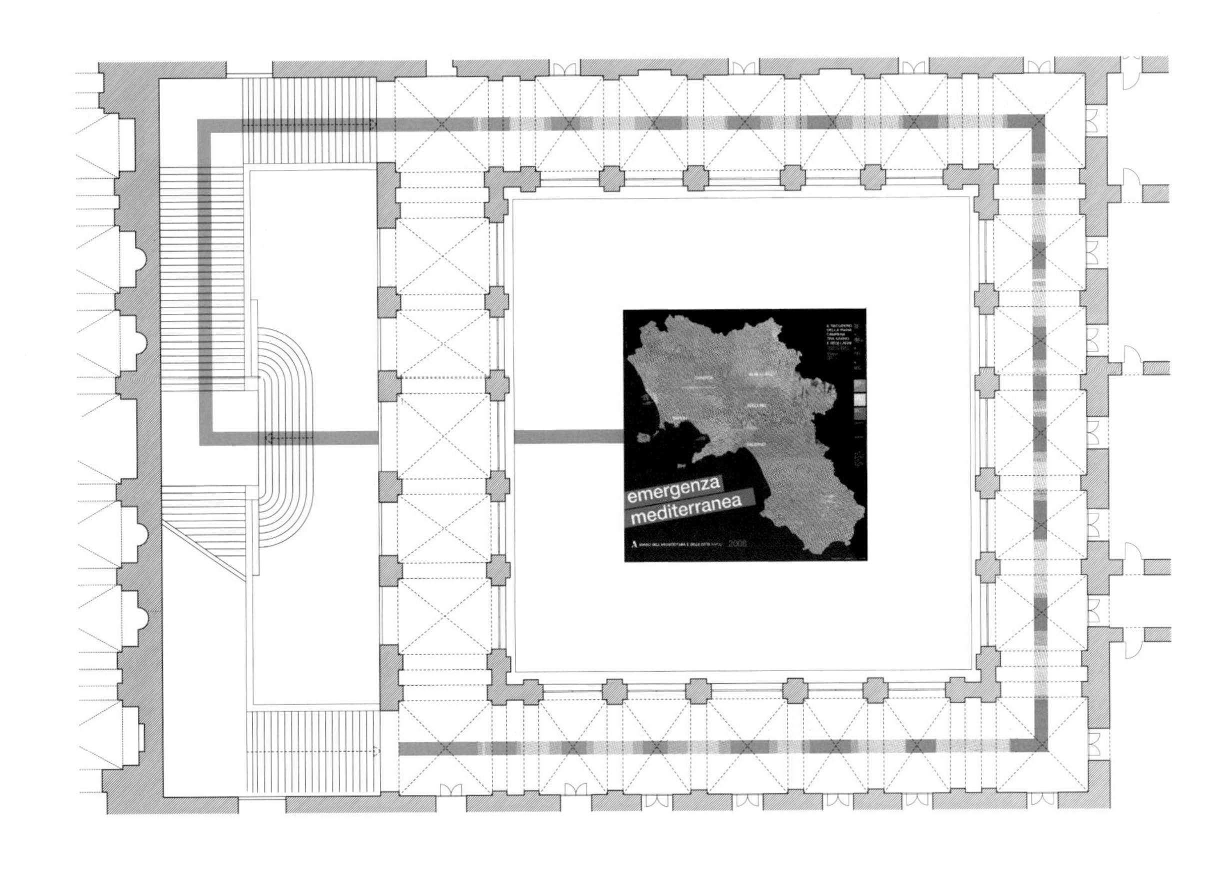

first floor plan
二层平面

0 5 10 m

A ANNALI DELL'ARCHITETTURA E DELLE CITTÀ NAPOLI 2008
ergenza
editerranea

PRO CAPITE DI

THE TABLE OF ARCHITECTU

Padua, Italy, 2008

地点 帕多瓦，意大利
项目 展览，装置
客户 Fondazione Barbara Cappochin
规划 2008年
建设 2008年

RE

the table of architecture views
预览建筑的展示桌

APPENDICES

附录

DIALOGUE

with Marco Casamonti

Four questions to tell a story

The question may be trite, but we would nevertheless like to ask you why you decided to become an architect?
At first my decision was completely casual; it was obviously one of those decisions one makes when one finishes high school and has to choose a faculty and to follow some inclination, which is not yet completely mature. However, before completing university I have acquired a more profound understanding of the fact that architecture, and the profession of architect, was to influence my life forever, becoming closely linked to every subsequent choice and prospect. After graduating and my first professional experiences, it was very clear to me that that discipline and body of knowledge, that I identified so much with, were the only way to express my desires and my youthful political aspiration. It gives one the possibility to improve, with one's work, the life and homes of people, to use one's culture and intellectual capacities to orient the decisions which influence the inhabited space and the landscape, a kind of microcosm within which it is possible to change, with one's work, a piece, even if tiny, in any case significant, of the surrounding scenario. The years have gone by but I am still convinced that the motivation is the same, to participate, we could say with "body and soul" but it would be more correct to say with one's awareness and knowledge, in the creation of small changes which may, in any case, concretely improve the conditions in which every individual plays his or her role in the community. It is a matter of a fascinating mathematical method which one studies at high school: "The principle of virtual works".
Moreover, even if I do not come from a family which had anything to do with architecture, I have certainly been influenced and attracted, from childhood, by visiting small building sites, as the construction of our family house, in which, as my mother remembers, I was the only one among my brothers who played with the workers, with trowels, lime and concrete, even as a child. More consciously, I have been fascinated with my father's activities, first as an art collector – my grandfather who was still an upholsterer knew Ottone Rosai and other Florentine painters – then as the founder of a contemporary art gallery and thus inevitably as an active player in the world of artistic expression and creativity. I remember some rare but extraordinary occasions when my father met artists as Vedova, Dorazio, Sebastian Matta, and while he discussed matters related to the organization of possible exhibitions and commercial aspects, I was more interested in the expressive possibilities of compositions and images which can certainly still be found, even if unconsciously, in my work today. I was fascinated by art and by artists due to their overall vision of the world, filtered and oriented in the images rendered in their works, something which can also be found, even if in a less direct manner, in architecture, an aspect which has always intrigued me, and still does.

Please tell us about your studies, and about the persons who have been important references to you.
On leaving school I spent more than a month in New York with Laura – then my study and life partner, and today also partner and mother of my children – whose brother lived in a somewhat dilapidated house in Queens. We decided to systematically visit all the museums in the city and to dedicate every day to a visit. Of all those explorations, I remember that I remained literally struck by the visit to the Guggenheim Museum by Frank Lloyd Wright; if truth be told I did not know and understand much about it, it was a matter of an unconscious and immature attraction, but it marked and reinforced my decision (Laura had never had any doubts) to enrol at the architecture faculty. Years later, I took another look at the incredible quantity of photographs I had taken in that period, and I was amazed to discover that there were very few close-ups, faces or typical snapshots, and that almost all of them were images of architectures and, more

对话马可·卡萨蒙帝

四个问题组成的故事

这个问题可能有点老生常谈，但是我们还是要问，你为什么决定成为一名建筑师？

开始，我的这个决定完全是随意的。我在高中毕业后必须选择一个专业，而且必须要有自己倾向的专业，虽然这种倾向还不是很成，但是，在我完成大学学业之前，我已经对建筑行业以及建筑师这个职业有了更深入的理解，认为它会影响我的一生，与我未来的选择和发展紧密相连。在毕业并获得第一次职业经验之后，我很清楚地认识到，我所熟悉的知识训练和知识储备会是我表达愿望和年轻的政治抱负的唯一途径。这个职业可以使人们通过自己的工作来提高人们的生活和居住条件，运用文化和智力来左右那些影响居住空间和居住景观的决定，可以通过这种微观的建筑行为来改变周遭的环境，即使这种影响是微不足道的，当然，有时，其影响也可能是巨大的。很多年过去了，但我的想法没有改变，我在去融入，融入人们的“身体和心灵”，或者更准确地说，用一些小小的变化来引起人们的关注和提高人们的知识，切切实实地提高每个人生活空间的条件。这就是我们在高中时所学到的一个有趣的数学原则：虚拟作品原则。另外，即使我并非来自一个与建筑有关的家庭，我在童年时代就受到了建筑的影响和吸引。在我家建房子时，我就会来到建筑现场，正如我妈妈回忆说，在我众多的兄弟中，我是唯一一个与建筑工人、泥铲、石灰和混凝土玩耍的孩子。另外，我对父亲的行为也非常感兴趣，他首先是一位艺术收藏家（我的祖父当时是一种装饰商人，认识奥托·罗塞和佛罗伦萨其他的画家），随后建立了一个现代艺术馆，并且必然地成为了艺术创新和艺术表现领域的积极参与者。我还记得，在极少的情况下，父亲会与诸如维多瓦、达拉兹奥、萨巴斯蒂安·马他等艺术家见面，在他们讨论与组织展览和商业活动相关的问题时，我们更加关注展览的表现效果，在我今天的作品中，虽然可能是无意识的，但仍然可以看到这种表现效果的痕迹。我对艺术和艺术家们表现世界的能力非常着迷，对他们作品的再现和表现能力着迷，这种表现能力虽然并未直接体现在我的建筑作品中，但一直到现在，仍然影响着我。

请谈谈你的研究状况，谈谈对你影响最大的人。

在离开学校的时候，我与劳拉（我研究和生活的伴侣，现在，她依然是我的伴侣， 也是我孩子的母亲）在纽约呆了一个多月，劳拉的哥哥在昆斯区拥有一栋基本荒废的房子。我们决定系统地参观纽约城所有的博物馆，每天参观一个。在所有的参观当中，我被弗兰克·罗伊德·怀特的古根海姆博物馆所深深吸引。说实话，我不理解也不明白其中的缘由，这是一种无意识的、不成型的吸引，但是，它却强化了我投身建筑事业的决心。多年之后，我再次欣赏当时拍摄的大量照片时，我很惊奇地发现，这些照片很少有特写、正面或典型的到此一游式的照片，它们大多属于建筑摄影，通常都是建筑物、基础设施、桥梁、车站、街道、博物馆或大楼。现在回顾起来，我可以说，这是我那一时期最重要的研究。在我80年代中期读大学时，我的转折点或者说是由一名学生转变成“献身建筑设计的人”的转型源自于阿道夫·纳塔里尼对内部装修的课程。我们当时已经完成其名为“设计与释放”的课程，但是我们只有在考试的时候对其进行讨论，因为多年以来，都有二百多个学生来听他的课程，因此，我们决定下一年度听他的另一个课程，主要考虑当时他已经转变了研究方向。在这些年里，阿道夫（我现在可以这样叫他，因为我们是朋友）是一位真正的领袖。各种重要的经验均来自于对过去的研究，他通过对原有大众建筑的修改和再解读，达到了自己研究生涯的顶峰。更重要的是，他的作品《石头的形象》也已由莲花出版社出版。对于学生来说，这本书无异于圣经。他的课堂尤如独角戏，象戏剧表演一样。他进入教室时会带一个录音机，课后他会将录音资料整理成书。他不会与任何人对话，但是，他的课程对于我们生活和作品来说依然发挥着重要的作用，不但具有说服力和感染力，也对建筑充满热爱。整年

in general, buildings and infrastructures, bridges, stations, streets, museums, skyscrapers. Looking back, I can say that was my first important study period. During my university years, in the mid-Eighties, the turning point or rather the transition from student to a "devotee of the subject" to use a somewhat bureaucratic expression, consisted of the lectures on interior decoration held by Adolfo Natalini. We had already completed his course in "design and relief" but we were only able to discuss it on the day of the exams because there were two hundred students attending every course in those years, and we therefore decided to attend another course held by him the following year, taking advantage of the fact that he had changed subject in the meanwhile. In those years Adolfo, as I can call him now that we are friends, was a true "guru". The radical experiences were by then relegated to the past, and he had reached the apex of his research activity through a critical revision and rereading of popular architecture and traditions, and moreover his booklet "Figures of Stone" had just been published by Lotus; it was a kind of bible for the students.
His lectures were true monologues, theatre performances. He entered the room with a recorder in his hand, which he used to transcribe a book he was preparing, he didn't say a word to anyone, but those dissertations were and remain extraordinary, fundamental for my life and work, persuasive and involving, an extolment of architecture. All through the year he left among applauses and general admiration, and so we decided to impress him with a project I still today consider wonderful, on the transformation of the Murate prisons in Florence, a project which was an act of devotion, also illustrated by a model in stone which I had made together with a friend who was a mosaicist and inlayer of semi-precious stones. Aldolfo Natalini was my first master. The opposite of Natalini – pupil of Savioli – was Loris Macci – pupil of Gamberini – he was the other side of the school, attentive, serious, a true academician, the profession and craft as an act of conscience and dedication, architecture as teamwork. We have graduated with him, with a project submitted to the competition for the Tokyo International Forum; the structure consultant was a personality much admired among those familiar with the study and discipline of the science of constructions: Salvatore Di Pasquale. I am not mentioning these persons because I like to list them, but because they have really been important for my training at a school which was a chaotic and savage multitude, but which gave you the sense of the university as a place where teaching was secondary to research and experimentation, something which is no longer the case today.
Another experience which has been fundamental to me has been my meetings, as fortunate as they were occasional, with Paolo Portoghesi: he had just been director of the Venice Biennial and as such customer of the Teatro del Mondo (Theatre of the World), then even president of the institution, in short he was, and still is to me, an extraordinary institution. My father commissioned a project from a Florentine architect who occasionally worked as collaborator of Portoghesi. I remember that I insisted on getting an opportunity to work for him, and to be able to participate in a project of his as a draughtsman, it was a matter of the renovation of the Marradi Pinery in Castiglioncello. I drew those plates with an extraordinary passion, I cleaned the office, I attached grids to tracing with a devotion which earned me a trip by car with Portoghesi: the architect sat in the driver's seat, Portoghesi sat beside him, and I behind. Portoghesi spoke about architects and architecture during the whole trip, in such a persuasive and fascinating manner that I was very impressed, just as I was amazed by the fact that he dictated to the recorder, off the cuff, a report which was perfect when transcribed, a kind of literary text. I understood that architecture was, beyond instinct and creativity, a scientific activity and the mirror of an intellectual transparency and clarity which certainly cannot be a secondary skill for a professional architect.

来，他都沉浸在掌声和羡慕之中，因此，我们决定用于个项目给他留下一些印象，至今我都认为这个项目非常精彩，这就是佛罗伦萨慕拉塔监狱的改造项目，倾尽了我的力量。我与一个镶嵌细工师朋友共同完成了石头模型的制作，他同时是一个准宝石镶嵌师。阿道夫•纳塔里尼是我眼中的第一位大师。与纳塔里尼（萨维奥利的学生）截然不同的一个人是劳瑞斯•马西（甘伯里尼的学生），他是学校里的另一类学者，专注、认真，真正的学院派，他们建筑行业看作良知与贡献，看作团队工作。我们在他的手中毕业，毕业设计就是一个参加东京国际论坛的项目，项目的结构顾问就是在研究和建筑科学领域倍受尊重的和熟知的萨尔瓦托•迪•帕斯奎尔。我之所以提及这些人，并不因为我喜欢他们，而是因为这些人对于我在学校接受的教育来说是至关重要的，而学校教育则是混乱无序的，但它可以告诉你研究和实验比教学更重要，而今天的情况已经大不相同了。
对于我来说，另一项重要的经历是与帕罗•波多盖西的几次相遇，真是幸运之极。他是上一届威尼斯双年展的主席世界剧院的顾客，同时也是该剧院的主席，简而言之，他依然是我的一位重要导师。我父亲从佛罗伦萨建筑师那里得到了一项代理项目。这位建筑师就经常与波多盖西合作。我还记得，我曾经坚持获得一个为他工作的机会，想作为制图员参与他的工程，即卡斯提格里昂塞罗的马拉第•派纳瑞改造工程。我以极大的热情制作图纸，打扫办公室，我努力以摹图并赢得了一次与波多盖西同车旅行的机会：建筑师坐在驾驶员的位置，波多盖西坐在他的旁边，我则坐在后面。波多盖西在整个旅程中都有讲有关建筑师和建筑的问题，其讲话方式迷人而又有说服力，给我留下了很深的印象。让我吃惊的是，如果将他的即兴发言录下来，其完美程度，不亚于一篇优美的文章。我明白，建筑并不仅仅是一种本能或创造，而是一种科学活动，是智慧通达的一面镜子，是一个专业建筑设计师的必备技能。
在众多直接的大师之中，还有很多间接的大师。奥瑞里奥•科泰西就是一个非常重要的角色，即使从感情的角度来讲，也是如此。他是一位非凡的人，一位非凡的建筑师。对于建筑学的教学和传授，对于建筑学的原则内容、价值和精髓，他的讲授和传授都甚称经典。由于性格内敛和文弱，科泰西没有创造历史，但是对于我来说，他在近年来的建筑史上是一个举足轻重的角色。这位默默无闻的潮流引领者出生于1931年，同一年，阿尔多•罗西、帕奥罗•波多盖西、吉尔吉奥•格雷西出生（真是人才辈出的年份）。在Ernesto Nathan Rogers主持卡萨贝拉研究中心的那些年，他一直在那里工作，现在他仍然是罗杰斯的学生，是二战以后借鉴意大利文化用于建筑的忠实履行者。我在22岁的时候就遇到了奥瑞里奥•科泰西，现在，在与他相识20年之后，他仍然是我学生和借鉴的楷模。在参加大学50年校庆的时候，我鬼使神差地听取了他首次为一年级学生开设的结构课程，并自此被他指定为助手。他让我做研究，使我明白了自己的研究必须超越普通的教育课程，他让我攻读博士，并在热那亚获得了博士学位。现在，我仍然在那里从事教学工作。

Among the few direct masters, there are a great many indirect ones; Aurelio Cortesi plays an important role, also from an affective point of view. An extraordinary man and architect. A true master in the sense of his vocation to teach and transmit architecture, its disciplinary contents, its value, its essence. Due to his reserved and timid character Cortesi will not become part of history, but to me he has played an important role in the history of architecture of the recent years, a silent and discreet protagonist who was born in 1931, the same year as Aldo Rossi, Paolo Portoghesi, Giorgio Grassi, (truly a vintage year). He has worked at the Casabella research centre in the years when it was directed by Ernesto Nathan Rogers, and still remains a pupil of Rogers, and a faithful divulger of a feeling which is the only certain and distinctive reference for the Italian culture of the years after World War II. I met Aurelio Cortesi when I was only 22 years old and now, after having known him for more than twenty years, he remains a continuous and constant reference. He appointed me as his assistant because he was surprised by the fact that, when attending the fifth year of university, I curiously followed some of his lectures in composition, first course, which he held for first-year students; he has made me study, making me understand that it was necessary to continue my education beyond the ordinary study program, he has made me enrol and participate in the Ph.D. which I then won at the architecture faculty of Genoa, where I still teach today.
He has been, to me and to my relationship with architecture, a kind of father, not just spiritually but concretely and factually. To help me in my university career, as he holds a regular professorship in Florence, he got a temporary assignment to teach architectural planning at the university of Genoa, where I was doing my Ph.D., and so every week, on Tuesdays, we drove from Florence at lunchtime when he had finished his lessons, reaching Genoa where we taught in the afternoon, he held fantastic lessons, I listened and learned, then in the evening I accompanied him once more, from Genoa to Parma. During the five or six hours of driving Aurelio Cortesi spoke uninterruptedly about architecture, he described the lives and personalities of the architects he had met and for whom he had worked since the Sixties, Gardella whom he knew and had frequented, Albini and Helg with whom he had worked for some years and above all Ernesto Nathan Rogers, his master and editor of “Casabella”; he had worked for the magazine together with personal friends as Giorgio Grassi or Aldo Rossi, to whom he was less close on an affective level but certainly not on a cultural one. Rather than mere journeys, these trips were very personal and private lessons of which memory is not just alive, but still fundamental today.
It has been a bit like attending university twice or thrice because this life (some said it was hard, but to me it was extraordinary and wonderful) continued for almost four years. When we arrived in Parma I often slept in his attic which served as library, and I usually did not manage to fall asleep because I was surrounded by all the magazines from the Fifties until then, “Domus”, “Casabella”, “Edilizia Popolare”, extremely rare issues of “Hinterland”, the magazine edited by Guido Canella, “Controspazio”, as well as an infinity of books which deprived me of my hours of sleep, but filled my head with a passion which gradually grew more and more visceral. I owe it to him that I am chief editor of “Area” today, because it was Cortesi who made me understand the importance of theoretical reflection and magazines as laboratories of ideas and fundamental instruments for the staging of a comparable critical dimension. By encouraging me to write and publish my reflections, he somehow “obliged” me to get to know some editorial offices, including “Area”.
At that time it was a matter of the magazine of a small publisher, Azzurra Editrice, directed by the mythical Mrs. Brivio who enthusiastically welcomed that youth who wanted to write for free and learn the ropes.

对于我和建筑之间的关系来说，他一直像一位父亲，并不只是精神上的指引而是实实在在的指引。由于他在佛罗伦萨拥有定期教职，为了帮助我的大学学业，他在热那亚大学争取了一个临时教授建筑规划的职位，我则在那里完成博士学位，因此，每周二，我们都会在他中午讲完课之后驾车从佛罗伦萨到热那亚，然后，下午，他在热那亚讲课，他的课非常精彩，我都会聆听和学习，然后，到了晚上，我再陪他从热那亚回到帕尔马。在五六个小时的路途上，奥瑞里奥•科泰西都不会断地讲述建筑方面的问题，他会讲述他自60年代以来接触过的建筑师的生活和性格，包括他所认识和熟悉的加德拉，以及共同工作多年的阿尔比尼和哈尔格，谈得最多的就是阿尼斯多 •纳斯 •罗杰，他的导师和“卡萨贝拉”的编辑；他与吉尔吉奥•格拉斯或阿尔多•罗西等朋友为该杂志工作，他们在文化上的亲近感比情感上的亲近感更多。我与他的这些旅程不仅仅是旅程，而是一种私人的授课，这种记忆仍然鲜活，并且到今天仍发挥着作用。

这样的生活好像读了3次或者4次大学，因为这种生活（有些人会说这种生活很累，但是对我来说，却是精彩的、有趣的）持续了几乎有4年时间。在回到帕尔马之后，我通常睡在他用做读书室的阁楼里，但是，我却经常不会立即入睡，因为我身边满是从50年代到当时的杂志，“Domus”、“Casabella”、“Edilizia Popolare”等，尤其是“Hinterland”杂志，该杂志由古伊多•卡尼拉主编，还有“**Controspazio**”。这里还有大量的书籍，阅读书籍占据了我大量的睡眠时间，但是却使我的头脑清晰而又充满活力。作为今天《Area》杂志的主编，我在他那里学到了很多，因为科泰西使我了解了理论研究的重要性，了解了杂志是思想的实验室，是发展重要思想的工具。他鼓励所写下并出版自己的理论，甚至“强迫”我认识一些编辑部，包括《Area》。当时，它还是一家小出版商旗下的杂志，由充满传奇色彩的布里维奥夫人主编，她热情地欢迎那些无偿写作并有很强求知欲的年轻人。

我的文章取得了一定的成功，随后，出版商为了应对危机请求我以较低的薪水从事该杂志的编辑工作。我接受了请求，条件是将杂志由季刊改为双月刊，可以自己选择编辑人员和顾问，并将编辑部移往佛罗伦萨。随后，杂志易手，被**Federico Motta Editore**买下，现在则由**Il Sole 24 ORE**拥有。当然，这是另外一回事儿了。

同时，对于我作为建筑师所接受的训练，**Federico Motta**也对我教诲良多。我遍游欧洲，与所有建筑著名出版商见面，企图挽救即将破产的杂志，因为原来的出版商和所有人无法保证杂志的成长，在此期间，我接触了**Federico Motta**。我吃了Skira及其所有者Vitta Zelman的闭门羹，她表示对建筑杂志没有兴趣。还有，我在Basel接触的**Birkhäuser**只对杂志表现出了少许的兴趣，直到我遇到**Federico Motta**，他告诉我，虽然他很欣赏我送给他的杂志，但是，他却没有兴趣出版《Area》杂志。但是，他向我提供了很有吸引力的出版社主管建筑方面的编辑职位，因为其前主管Pier **Luigi Nicolin**（《莲花》的主编）将要去**Electa**。面对这一要求，虽然我有一些不知所措甚至诚惶诚恐（因为Nicolin的书籍非常精美），但是我还是接受了邀请，并且提出了一些新的编辑规划，另外，针对意大利建筑争鸣中的建筑、制图等问题出版了一些新书，以此希望能够从事自己熟悉的领域，试图在不冒文化风险的前提下做出一些改革尝试。自然而然地，第一本书选择了有关阿道夫·罗西的有关绘图和制图的专题，该书是与他的儿子**Fausto**和他的事务所合作完成的。**Federico Motta**允许我学习一些原来不了解的门类，并且将其与建筑师的身份结合起来，换句话说，就是将其结合到建筑和创新当中。数个月之后，**Motta**也开始了《Area》的出版，其新的平面设计由**A G Fronzoni**提出创意，在数次的会面当中，他对平面设计的热情感染了我，这种感染一直持续到了现在。当然，我也不应当忘记在我生命中同样非常重要的其他人，包括《Area》杂志编辑委员会的所有成员及其指导和在杂志社实习的学生，以及我读博士时的一些同学还有杂志社编辑部的工作人员。不付报酬、自愿的工作，仅靠热情和参与的愿望来支撑，对于我来说，繁杂却很有趣，我希望，对其他人来说，也是如此。他们中的一些人，与我、我的生活和学习有着非常紧密

My articles enjoyed a certain success and following and the publisher, facing a crisis, asked me to edit the magazine for an objectively low fee. I accepted, on the condition that I was given free reins to transform the magazine from quarterly to bimonthly, choose the editorial staff and the consultants, and move the editorial office to Florence.
Then the magazine has changed hands, it has been acquired by Federico Motta Editore and is now owned by Il Sole 24 ORE, but that is another story.
However, also Federico Motta has meant a lot for me, and for my training as an architect. I contacted him after having travelled all over Europe, to see all the leading architecture publishers in an attempt to save the magazine from bankruptcy, in any case from an impossible growth which the old publisher and owner could not guarantee.
I only got doors slammed in my face: from Skira and its owner Vitta Zelman who expressed her lack of interest in architecture magazines, to the moderate enthusiasm of Birkhäuser whom I met in Basel, until Federico Motta called to tell me that even though he appreciated the magazines I had sent him he had no intention of publishing "Area". However, he offered me a very prestigious position as editorial director of the publishing house's architecture sector, as its former head Pier Luigi Nicolin, being editor of Lotus, was in any case bound to Electa. Although I was scared, or perhaps I should rather say terrified, by the proposal – the books created by Nicolin were very refined – I accepted, and suggested a new editorial line which met with success, also in commercial terms, by inventing some new book chains, on building categories, on the drawings and sketches of the leading players in the Italian debate, seeking to move in a territory I was familiar with, trying to be innovative without taking too many risks on a cultural level. The first book was, intuitively, one on drawings and sketches dedicated to the work of Aldo Rossi, prepared in collaboration with his son Fausto and his firm. Federico Motta allowed me to learn a profession which I objectively did not know, and to do so as an architect, and in other words to conceive it in terms of design and creativity. After some months Motta also began to publish "Area", with a new graphic design ideated by A G Fronzoni, who in a few meetings infected me with a passion for graphic design which I still have. But it would be ungenerous to forget other figures who have been extremely important to me, as every member of the editorial committee of "Area", and their students who formed and who still participate in, together with some fellow students from my Ph.D. course, the editorial staff of the magazine. A free and voluntary work, only motivated by an enthusiasm and desire to participate, arduous but interesting to me and, I hope, to the others. Some of them have been especially close to me, my life and my studies: Alessandro Anselmi, Augusto Romano Burelli, the lamented Pasquale Culotta, Claudio D'Amato and Franz Prati, all of them extraordinary persons in their own ways; I met them through a former pupil of Aureilo Cortesi who was then chief editor of the "Materia" magazine. I owe a lot to Claudio D'Amato in terms of ethics and school, Claudio has taught me to recognize the value of people and things; thanks to his intellectual honesty, he made me understand the value and importance of research within the university allowing me to become associated professor at a very young age, only two years after having won the Ph.D. competition at the University of Florence.
As to architects, I have learned a lot from three persons: Franco Purini, Arata Isozaki and Rafael Moneo. The first has been a virtual professor to me, he has taught me a lot through his writings and works. The most worn-out book in my library is beyond any doubt his "Seven Landscapes" published by Lotus, I have read and browsed it dozens of times, while his critical intelligence and profound and exuberant analytic skills have been like an energy, nurturing me with vehemence. I recall a memorable interview-meeting for my Ph.D., I cannot forget his intellectual generosity, prodigious with advice and references, a generosity with which I completely identify. Today I consider him a friend, someone who

的联系：Alessandro Anselmi、Augusto Romano Burelli、the lamented Pasquale Culotta、Claudio D'Amato以及Franz Prati，他们都有各自的领域非常出色。我是通过Aureilo Cortesi以前的一个学生接触到他们的，Aureilo Cortesi现在是Materia杂志的主编。在品德和学校教育方面，我在Claudio D'Amato身上学到了很多东西。他教我如果认识人和思想的价值，我非常感谢他在知识上的慷慨，他使我明白了大学研究的价值和重要性，使我在很小的年纪就成为了副教授，当时，我刚刚在佛罗伦萨大学获得博士学位两年。对于建筑师，我从下面三个人身上学到了很多东西：Franco Purini、Arata Isozaki和Rafael Moneo。Franco Purini可以说是我虚拟的导师，他的文字和作品使我受益匪浅。我书架上看得最多的书就是莲花出版社出版的《Seven Landscapes》，我曾经数十次地精读和泛读过这本书，他批评的眼光、深沉而又完美的分析技巧，尤如一种能量，一直滋养着我。我想起了我博士学位的一次答辩会议，无法忘记他在知识上的慷慨，给我提了很多建议，我完全认识到了他的慷慨。现在，我把他看作朋友，虽然我们很少有机会见面，只是虚拟上的朋友，但是，对于我来说，他非常亲切，这种感觉无法用语言来形容，我对他感觉非常亲切，非常尊敬他。我曾经将一篇专题论文献给Arata Isozaki。由于他一直参与有关现代建筑问题各个阶段的争鸣，我希望我的研究能得到他的认可。在意大利与他数次会面之后，我去了东京，在他的工作室呆了一周，那段经历以及与他的交谈，对于我来说是很重要的、有意义的。我已经明白了所谓"研究"的含义，它是独立于文字的，只专注于风格、事务的本质和精髓。对于Rafael Moneo，我与他见面的机会很多，我曾经几次去马德里，我们曾经多次合作设计项目（当然都是以老师和学生的关系），如佛罗伦萨原军事面包店区域的项目（现已废弃），以及威尼斯双年展电影宫的第二次竞赛项目，这个项目，我们得了第二名。无论结果如何，这都是非常难忘的经历。Moneo当然也是我的榜样，对于我作为一名建筑师来说，由于他对建筑行业的贡献永远值得学习，所以，与他一起参观建筑现场是非常重要的。虽然是一名出色的设计者，Moneo也会教我建筑现场的重要性，告诉我年轻的合作者必须经常在现场，在建筑的现场生活和工作，直到完工。我之所以尊敬他，不是因为他是令我印象深刻的哈佛学院的主任，也不是因为他是Pritzker奖的获得者或者其他的无数荣誉，而是因为他的意志力以及对项目的贡献。在研究过程中不断追求卓越，我认为，这是他工作中很自然的一部分，同时也是他长时间不断坚持工作的结果，他在Calle Cinca工作室中大量的模型及其相关细节就是证明。最后但并非最不重要的，我还要提及那些与我在同一个工作室工作的同事们：Laura Andreini、Silvia Fabi、Massimiliano Giberti、Giovanni Polazzi。他们不只是我的榜样，也是我的朋友和兄弟，他们已经是我生命的一部分；其中一些人已经成为我的左膀右臂，一些已经成为我道德良知的一部分，一些已经融入我的心灵之中。总之，我必须认可：他们是一股自然的力量。

is dear to me even though, to some extent, virtually, as we have had few opportunities to meet; in any case he is someone whom, it is hard to explain with words, I feel close to; I have a profound esteem for him. I have dedicated a monographic issue to Arata Isozaki, a deed of recognition of the value of a research which has always intrigued me due to the flexibility, the convinced participation in every phase of the contemporary architectural debate which he has followed in every part, always playing an active role during the different phases of his life. On that occasion, after some sporadic meetings in Italy, I went to Tokyo and spent a week in his studio, an experience and dialogue that has been fundamental, and enthusing, to me. I have understood the meaning of a research work which goes beyond the calligraphy, which focuses, independently of styles, on the substance of things, on their essence. With Rafael Moneo the opportunities to meet have been more numerous and continuous, I have travelled to Madrid on several occasions, we have designed a number of projects together, obviously on the basis of a pupil-master relationship, as the project for the area of the former military bakery in Florence, which was then abandoned, and the project for the second competition for the Movie Palace of the Venice Biennial, where we came second. Regardless of the results, these have been unforgettable experiences. Moneo is certainly a reference and the visits to building sites made in his company have been very important to my life as an architect, due to his dedication to a profession where you never stop to learn. Although an extraordinary intellectual, Moneo has taught me the importance of the building site and the necessity of always having young collaborators at the site, who live there and work on the construction, never abandoning it until it is finished. It is not the fact that he has been director of the Harvard school which has impressed me, nor that he has won the Pritzker or his countless other recognitions, but his willpower and total dedication to his projects.
The extraordinary research for sublimeness, which I believed to be a natural aspect of his works, and which on the contrary was and is the result of a continuous and insistent work, as witnessed by the numerous models of a building or its details which his studio in Calle Cinca was filled with.
Last but not least, my references are the people who share the studio with me, Laura Andreini, Silvia Fabi, Massimiliano Giberti, Giovanni Polazzi; they are more than references, friends or siblings, they are part of me; someone my hands, someone else my conscience, someone my heart, all together, I must recognize it: a force of nature.

Can you tell us about your first work experiences and projects, and the role which research has played in this context?
Research and design are inseparable, they are one and the same thing. A project is a work of research, which is always developed through a project, an idea, a mental construction which precedes the commencement of the research as such. One does not begin an experiment just to "find" something, one first of all begins, precisely, with a project, a research project, which then makes it possible to identify possible solutions. At the same time one does not design just to build, but to experiment the result of a research which reflects one's knowledge and cultural approach to architecture and how we inhabit it. I believe that I have expressed these concepts, with my life, my work, my personal trajectory; from the beginning I have instilled, in my projects, the effort of research, the conviction that architecture is more than merely an act of superimposition of one's personal style on a place or a building; rather, as Adolfo Natalini has written, we can imagine our work as that of a water-diviner who, holding a piece of wood – a pencil in the case of the architect – searches for water veins in the earth. However, I am not completely convinced by a concept of a "Michelangeloesque" kind, according to which a sculpture is already contained within the marble block from which the artist removes

你能为我们讲一讲你最初的工作经验和最初所从事的项目，以及研究工作在其中的作用吗?

研究工作与设计工作是密不可分的，两者是一体的。一个项目就是一项研究工作，而研究则通过项目、思想和思维的构建来发源。人们不是通过实验来“发现”某些事情，而是通过一个项目来开展研究，发现解决方案。同样，人们的设计也并只是为了实物建筑，而是对研究结果进行实验，检验一个人在建筑上的知识和文化体验以及如何将其付诸实践。我认为，我已经通过我的生活、工作以及人生轨迹表达了上述思想。从一开始，我就一直在我的项目中发挥研究工作的力量。我认为，建筑不仅仅是将一个人的风格注入到某一地点或建筑物之上，而是如阿道夫·纳塔里尼所写的那样，我们可以假设自己是一个探测水源的人，手持一根木棍（就是建筑设计制图时的笔），在地球的表面寻找水道。但是，我并不认同“米开朗基罗式”的说法，即艺术家的工作只是将大理石的表面剥开，发掘大理石内部的美，这就等于抹杀了一个项目作为智慧产品的价值，抹杀了其表达意义。据此，我认为，建筑师的工作应当是一种对某个地点和项目改造活动的挖掘、描述和识别。我相信，这种做法可以在我最初的作品中得到体现。在我最初的重要作品当中，贝加莫迪斯科舞厅（很不幸，现已不存）见证了对于建筑价值的研究，而且不仅仅体现在形式上。我们前期的经验主要集中于内部装饰，我们从未在严格意义上重要这项工作。我们是作为城市空间来对这些空间进行设计的。我们将这些装饰作为建筑的组成、石头的堆砌、墙面或房屋的组合。相反，我们在贝加莫，在这样一个被棚户和工业建筑所主宰的城市，有机会首次体验了城市空间、城市最复杂的部分、边缘的部分、郊区以及城市的扩展部分。我们接受了这个挑战，开始处理新建筑与城市外缘、高速公路、广告牌以及前方长满树木的小山之间的关系，并做好改造工作（那是一个70年代设计师设计的普通混凝土建筑）。我们被曾经的工业记忆所深深吸引，希望使该建筑发挥新的角色，即成为一个娱乐场，承载我们所熟悉的艺术世界所带来的感觉：Lucio Fontana的入口设计；Jannis Kounellis对钢铁的运用；Richard Serra所创造的铁锈的斑驳效果；阿曼为在入口处制造地板而设计的堆积效果，即很多堆积在地下机板上的瓶子碎片。建筑现场的工作经验教会了我如何管理大型工程，对我来说是一种工作方式的积累，甚至有些时候，晚上也会在现场。我们立即认识到，在第六届威尼斯建筑双年展上，我们提出了一项独一无二的作品，已经得到很多评论和出版物的认可，并且在很多邀请要求我们展示建筑正面的那部分。

the superfluous material, because this is tantamount to cancelling the value of the project as intellectual product, as narration. I think one should capture, from that suggestion, the idea of the architect's work as a discovery, a retracing and identification of the potentials of a place or theme with respect to the modifications introduced by the project. And I believe this approach can be found in my work from the very beginning; among our first important projects, the big discotheque in Bergamo, which unfortunately no longer exists, bears witness to this research for values, not only figurative, within a work of architecture. Our few previous experiences had been limited to interiors, a type of approach we have never considered in the strict sense; we designed those spaces as urban spaces and those interiors as fragments of buildings, splinters of stones, pieces of walls or houses. Vice versa in Bergamo we were, for the first time, given an opportunity to actually measure swords with the urban space, the most complex part of that city, the marginal one, the outskirts and their urban sprawl, with a superimposition on a tissue and environment dominated by sheds, by industrial buildings. We have accepted that challenge and begun to work on the relationship between the new building, in the process of transformation – it was a matter of a banal concrete structure built by craftsmen in the Seventies – and the surrounding landscape, trying to find a path halfway between a continuity – the image of the city outskirts, the highway traffic, the billboards, and a more ample context of the forest-clad hills in front. We were attracted by the traces of the memory of the old industrial function and the need to express the new role played by the building, an entertainment venue, conveying the sense of a figuration originating from our particular familiarity with the world of the figurative arts: Lucio Fontana, for the openings, Jannis Kounellis, for the use of steel, Richard Serra for the fascinating effects of rusted surfaces, Arman's accumulations, used to create the floor of the entrance area, a heap of bottle shards placed underneath glass panes. That building site has taught us to manage a large project, it has represented the apprenticeship of a way to work, also by night, which unfortunately still continues. We have immediately realized that we had conceived a singular work, something which has been confirmed by the positive reviews, the numerous publications, the invitation to participate, with a fragment of that façade, in the VI Architecture Exhibition of the Venice Biennial.

What has been the turning point in your professional career?
We have for many years been considering the last project as the best, as the one marking a turning point; vice versa the "turning point", meaning success and recognition of one's work on an intellectual and cultural level, is necessarily a slow and steady process which is not easily perceived on a short-term basis. What has changed, between the first and second decade of activity, is the possibilities of the work offers, which were sporadic, laboured and sought-after in the first years, and more numerous and never requested in the last decade. We have realized that we must have done something right when we no longer needed to look for assignments, which on the contrary have begun to arrive with a certain regularity, forcing us to constantly expand our operational structure. We have begun to select the proposals on the basis of the quality of the customers, and fortunately, as we today work in various parts of the world, from Italy to China, to North Africa and the Arab Emirates, we have not suffered much from the crisis which has affected the building sector in the last year, because our strategic diversification and resources have come to our aid. On the contrary, a wanted and sought-after change is associated with the dimension of the studio, which is conceived as a workshop or atelier, a research laboratory which, as it could not acquire an industrial dimension, once we had reached forty employees – the most

你事业的转折点是什么？

多年来，我们一直将最后一个项目看作最好的，将其作为一个转折点；相反，“转折点”一词意味着成功，意味着作品在知识界和文化界得到认可，这必定是一个缓慢、稳定的过程，不可能在短期内实现。在第一个十年和第二个十年之间，变化的，只是工作邀请的可能性。这些工作邀请只是零星的、吃力的，并且在最初的几年里还很受欢迎，在后一个十年里，则数量更多并且从未提出过要求。我们认识到，在我们不必为项目发愁的时候，我们必须做该做的事情。我们的项目来源逐渐稳定，扩大了我们的工作范围。我们开始依据客户的质量搜集建议，幸运的是，我们今天的工作领域遍及世界，从意大利到中国，再到北非和阿联酋，我们不再受建筑行业普遍面临的危机之苦，因为我们的战略是多元化的，来源广泛。相反，我们所希望并且受到欢迎的变革与工作室的工作范围相关联。我们的工作室被看成了一个车间或工作间或一个研究实验室。随着我们的雇员达到40人（佛罗伦萨同行中员工人数最多的事务所），它就会出现一种工业企业的状态，分成各个部门，分成很多房间，分布在不同的城市当中，有时各自之间距离很远，当然，很多情况下，规模并不大。当然，工作室永远也不会像一家公司。我还记得福斯特公司的经理多年前邀请我参加一个会议时的警告，他的事务所与Ove Arup刚刚赢得了佛罗伦萨高速火车站的一个项目，我们也参与竞争，但是很不幸，他们只选择当地的事务所进行合作。虽然我很疑惑，并有一种被欺骗的感觉，然后我去了伦敦。相反，事实证明，这段经验开阔了眼界，并且是非常重要的，我知道了作为一名建筑师永远不要去做什么。例如，成立一家看起来像银行的公司，有穿着制服的秘书和司机，有休息室，不但有很多设计师，也有很多经理，其中一人自信地告诉我，他通过一家我们所不了解的市场调查公司选定了我们。寒暄之后，在发现我们无法提出设想甚至连一个门把手都无法设计的时候，我谢过之后，离开了。我宁愿要Calle Cinca的镶木地板，虽然有些不合时宜，但很人性化，很有文化感，很新鲜，上面很斑驳，写满了历史和记忆，在泰晤士河上闪闪发光。当我们认识到事务所在成长，需要更多空间时，我经常会准确地说出这个故事。我将这种感觉与伦左皮亚诺的话进行对比，他在一次很不正式的谈话中透露，他很喜欢自己在热那亚的工作室，还有在巴黎的工作室，因为他们不喜欢他增加员工的人数和工程的数量，正是由于这种限制，他才能够更好地控制事务所的项目。Piano所说的话，即他不能管理像福斯特那样的事务所（超过600名员工），可能会令我们发笑，但是，它告诉了我们这个行业作为一家具有创新能力的事务所的人数上限所在。当你以一名艺术家或作家而工作，

the studio in Florence could hold – has grown by fragments and separate rooms, scattered across different cities, sometimes distant from one another and in any case never very big. The studio has therefore never come to resemble a company. I remember, as a warning, when a manager of the Foster firm called me for a meeting some years ago. Their firm, along with that of Ove Arup, had just won the competition for the high speed train station in Florence; we had participated as runners-up, but without any luck, and they were looking for a local firm with which to collaborate. Intrigued, even if somewhat sceptical, I went to London. Vice versa, the experience proved clarifying and important, because I understood exactly what I would never want to do as an architect. For instance create a firm which looks like a bank, with secretaries in uniform and drivers, waiting rooms, many designers but also many managers, one of whom somewhat proudly told me that he had chosen us through a marketing survey entrusted to I don't know what company, which had chosen us. After a few civilities, having ascertained that we could not have ideated and designed even a handrail, I thanked and left. I certainly prefer the worn parquet of Calle Cinca, somewhat untrendy but human, cultured, refined, full of scratches, histories and memories, to the glittering glazed façades on the Thames. I often tell this episode precisely because when we realized that the firm was growing and we needed more space, I compared those sensations with the words of Renzo Piano who, during a quite informal conversation, confided that he was fond of his studio in Genoa and of that in Paris because they did not allow him to increase the number of persons and thus of the projects and that he could, because of those limits, maintain a better control on the projects the firm worked on. Piano's assertion, that he would not have managed to work in a dimension as that of Foster – more than 600 employees – may perhaps make us smile, but it is indicative of a dimension, that of creative workshop, which cannot be neglected in our profession.
You reach your professional turning point when you manage to work as an artist or writer rather than as a technician, with the completely Loosian awareness of being little more than a carpenter. This is why there is nothing sensational about our sought-after international dimension, which is by no means a sign of success but rather indication of a nomadic attitude which make us go to the places where things really happen, at least as far as our work is concerned.
We have a small studio in Beijing because we want to be in China, and familiarize with a such a different reality, which is undergoing such an extraordinary transformation. We have a room in Dubai because we want to measure swords with a hyper-place that is gawky, postmodern and in any case unique in the international scenario, a kind of ebullient landscape where land becomes sea and water land, where we are designing a skyscraper and the anthropization of a heap of sand in the middle of nothing. We are working in Albania, where we are building the Tirana tower, and in Libya where we are designing the new Architecture Faculty of Tripoli. We hope to be able to go to India soon, to Bombay and also to Central Africa, where architecture is really needed, and not just someone's whim or status symbol. This is the true professional turning point, to be able to move freely in the world with one's briefcase and convictions, with the curiosity and intensity which certainly animated "Corbu" in his frenetic movements from Germany to France, from Moscow to Algiers, to the epic of Chandigarh. He is an unattainable genius, while we are perhaps just travellers, but that's fine with us.

Extract from the interview made by the Commissione Giovani dell'Ordine degli Architetti PPC di Lecco.

而不是技术工人而工作的时候，你才会真正迎来事业的转折点，技术工人仅仅比木匠强那么一点点。这就是为什么我们不应该欢迎国际化的问题，国际化不是一种成功，而是一种游荡的心态，使我们真正到事情出现的地点，只要与我们的工作相关。我们在北京有一个小的工作室，因为我们希望到中国工作，并且对这种不同的现实情况很熟悉，中国正处于巨大的转型之中。我们在迪拜也有一个工作室，因为我们想在这个古老、滞后而又在世界中独一无二的地方创造奇迹。那是一个热情的地方，土地变成了海洋或海洋中的土地。我们在那里设计了一栋摩天大楼，在一片荒芜中堆起沙堆。我们在阿尔巴尼亚工作，我们在那里建设提拉那塔，在利比亚，我们在的黎波里设计建筑学院。我们也希望很快就能打入印度市场，去孟买或中非，那里很需要建筑，而是仅仅是一个标志性的建筑。这才是真正的职业转折点，带着自己的箱子和信心自由地走遍世界，像柯布西耶一样，带着好奇和热情从德国到法国，从莫斯科到阿尔及尔，再寻找昌迪加尔的史诗。他是一位无法企及的天才，所以，我们可能只是旅行者，但这就足够了。

节选自意大利莱科青年建筑师协会的采访。

BIOGRAPHY

Laura Andreini (born 17 June 1964 in Florence), Marco Casamonti (born 13 April 1965 in Florence) and Giovanni Polazzi (born 19 December 1959 in Florence) are architects who graduated from the Faculty of Architecture at the University of Florence. In 1988, in Florence, they founded Studio Archea, joined by Silvia Fabi in 1999, Gianna Parisse from 2001 to 2008 and Massimiliano Giberti in 2002.
In addition to their principle work in architectural design on projects on every scale, from the object to the building to the urban plan, each studio associate is also involved in intensive teaching and research work in the discipline of architectural design in faculties of architecture throughout Italy.
Over the years, these pursuits have been supported by in-depth critical study and writing on architectural issues, seen in essays and articles published in books and magazines in Italy and abroad, as well as in the organization and production of exhibitions, events and workshops related to design.
In addition, since 1997, Marco Casamonti has been editor-in-chief of the international architecture magazine "Area"; since 1999, he has been co-editor with Paolo Portoghesi of the magazine "Materia"; also since 1999, he has been technical editor of the architecture section of Motta Architettura of Il Sole 24 ORE Group; and from 2006 to 2008, he was technical director of the "Annali dell'Architettura e delle città, Napoli" foundation, for which he and the studio organized and coordinated events in all three years of the event.
The studio has designed numerous architectural projects, which have been published in books and major international magazines (including "Abitare", "Casabella", "Domus", "l'Arca", "In Italia", "A&V", "Disegno Interior", "AIT", "Detail", "a-+u") and have been chosen for major architecture exhibitions. In 1996, with its design for an entertainment center in Curno, the studio was invited to participate in the Italian section of the 6th International Architecture Exhibition at the Venice Biennale; in 2002, with the Leffe House, it was chosen for the international stone architecture exhibition and an Italian architecture exhibition in Tokyo called "From Futurism to a Possible Future"; and in 2008, its design for a library in Nembro was chosen for the exhibition organized by the Design Museum in London.
Studio Archea has participated in major national and international architecture competitions and consultations, earning much recognition and many awards. In 1998, the studio won the competition for an administrative and commercial center in Calenzano, Florence; in 1999, it won third place for the new Faculty of Architecture location in Venice; in 2000, second place for the Italian Space Agency's headquarters in Rome; in 2001 and 2002, it was a finalist for the high speed rail stations in Turin and Florence; in 2002, it was invited by the Immobiliare Novoli real estate agency, under the technical direction of Aimaro Isola and Francesco Dal Co, to be among the nine Italian groups chosen to design a block in the former Fiat areas in Florence; in 2003, it won a competition to expand the port of Savona and a competition for the design, renovation and adaptation to regulations of Michelangelo campground in Florence.

发展历程

Laura Andreini（1964年6月17日生于佛罗伦萨）、Marco Casamonti（1965年4月13日生于佛罗伦萨）和Giovanni Polazzi（1959年12月19日生于佛罗伦萨）都是毕业于佛罗伦萨大学建筑系的建筑师。1988年，他们在佛罗伦萨创建了Studio Archea建筑师事务所，随后，Silvia Fabi于1999年加入，Gianna Parisse在2001至2008年间加入，Massimiliano Giberti于2002年加入。
每位事务所合伙人都在从事各自的项目建筑设计这一主要工作，而这些项目涉及了各种规模，其范围从建筑物到城市规划，此外，他们还参与了遍及意大利各地的建筑学系的建筑设计学科的大量教学和研究工作。多年以来，他们的不懈追求总能得到他们所开展的深刻评论研究和建筑主题写作的支持，这些可见于在意大利和国外出版的书籍和杂志上所刊载的文章和论文，并体现在与设计相关的展览、活动和研讨会的组织和作品设计工作中。
此外，自1997年起，Marco Casamonti一直担任着国际性建筑杂志“Area”的主编；而且自1999年起，他一直担任着Il Sole 24 ORE Group出版集团旗下Motta Architettura公司的建筑学科的技术编辑；从2006年－2008年，他担任了“Annali dell'Architettura e delle città, Napoli”基金会的技术编辑，此基金会在这3年中举办的所有活动均由他本人和本事务所组织和协调。
本事务所设计了数量众多的建筑项目，这些项目刊载于书籍和主要的国际杂志上（包括“Abitare”，“Casabella”，“Domus”，“l'Arca”，“In Italia”，“A&V”，“Disegno Interior”，“AIT”，“Detail”，“a+u”）并入选了各大建筑展览。1996年，因其对位于Curno的一个娱乐中心的设计方案，事务所受邀参加了第6届威尼斯双年展国际建筑艺术展的意大利部分的展览；2002年，事务所以Leffe House项目入选了国际石材建筑展和一个在东京举办的“从未来主义到可能未来”的意大利建筑艺术展；2008年，事务所的Nembro图书馆的设计方案入选了由伦敦设计博物馆组织的展览。
Studio Archea一直在参与重大的国内和国际建筑竞赛和咨询工作，赢得了大量赞誉和许多奖项。1998年，事务所赢得了位于佛罗伦萨Calenzano的一座行政和商业中心的设计竞赛；1999年，在位于威尼斯的新建筑学系项目设计竞赛中赢得了第三名；2000年，在罗马意大利空间署总部设计竞赛中赢得第二名；2001年和2002年，成为了都灵和佛罗伦萨高速铁路车站设计竞赛的决赛者；2002年，受Immobiliare Novoli房地产公司之邀，在Aimaro Isola and Francesco Dal公司的技术指导下，跻身于入选设计佛罗伦萨前菲亚特厂区街区设计的9家意大利集团之一；2003年，赢得了佛罗伦萨的Savona港口扩建项目设计竞赛以及Michelangelo宿营地的设计、翻新和法规适应设计竞赛。
2004年，与建筑师Rafael Moneo合作，事务所受邀参与了威尼斯新Palazzo del Cinema项目的设计竞赛。2005年，受邀参与了地拉那新城市中心总体规划当中十大塔楼之一的设计竞赛并赢得了第一名。
2005年11月，赢得了受Pirelli Re和Morgan Stanley之邀参加的国际设计竞赛－与Michael Maltzan

In 2004, with architect Rafael Moneo, it was invited to take part in the competition for the new Palazzo del Cinema in Venice. In 2005, it participated in a competition by invitation to design one of the ten towers planned for the master plan for the new center of Tirana, winning first place.
In November 2005, it won the international competition by invitation put on by Pirelli Re and Morgan Stanley – in a tie with Michael Maltzan Architecture – for the former Ansaldo area in Milan in the Grande Bicocca area.
In 2007, it participated in the competition for the final design and execution of projects for the Music and Culture Park, winning 2nd place. The studio also participated in the "Lonely Living" extranext event at the 8th International Architecture Exhibition of the Venice Biennale; at the Biennale in 2003, it was invited to design the entrance for the 50th Biennale for figurative arts with the traveling piece, "The Cord", designed by Archea Associati with C+S.
In 2007, the new municipal library of Nembro and the Albatros campground in San Vincenzo were opened; in 2008 the Edison bookstore was opened within the former Lazzeri Theatre in Livorno; and in 2009, the new city hall of Merate and the municipal library and auditorium in Curno were opened.
Current projects under construction include a large square with services in Merate near Lecco; an educational social center in Seregno, a residential and commercial center in Tavarnuzze, near Florence, the new Cantina Antinori winery in Bargino, San Casciano Val di Pesa, the new facility and offices of Perfetti van Melle in Lainate, Milan; the Tower of the Arts in Milan, the Borgo Arnolfo new residential and commercial center (former Alberti hospital) in San Giovanni Valdarno, Florence, the Tower of Tirana, Albania, the KPM Tower and the "Meravigliosa Island" in "The World" in Dubai, UAE.
In industrial design, the studio has produced numerous patents for the design of products and parts for lighting and construction materials, including the design of reconstituted terracotta in a joint venture with two leading companies in the industry.
These projects involve intense collaboration with the construction materials industry in order to implement a practical, effective collaboration between the worlds of design and manufacturing. This gives the studio the opportunity to pursue experiments and research based on the use of materials.
In the studio's offices in Florence, Rome, Milan, Beijing, Dubai and São Paulo over eighty people work, including architects, designers and graphic designers, as well as secretaries, administrators and public relation managers, from universities and countries around the world.

Architecture事务所打成平手,竞赛项目是Grande Bicocca地区米兰的前Ansaldo区。在2007年，参与了“音乐文化公园”最终设计和项目实施的设计竞赛，赢得了第二名。事务所还参加了威尼斯双年展第8届国际建筑艺术展的“Lonely Living”展外活动；在2003年的双年展，受邀设计了第50届旅游品人物艺术双年展的入口，即“The Cord”，由Archea Associati联合C+S设计。

2007年，Nembro新市政图书馆和San Vincenzo的Albatros宿营地开放；2008年，Edision书店在Livorno的前Lazzeri剧院内开放；在2009年，Merate新市政厅和市政图书馆和Curno的礼堂开放。目前在建的项目包括在Lecco附近的Merate的大型广场及服务设施；一座位于Seregno的教育社会中心、位于佛罗伦萨附近Tavarnuzzea的一座住宅和商业中心、位于San Casciano Val di 的Bargino的一座新Cantina Antinori酿酒厂、位于米兰Lainate的新Perfetti van Melle设施和办公室；位于米兰的艺术之塔（Tower of the Arts）、位于佛罗伦萨San Giovanni Valdarno的Borgo Arnolfo新住宅和商业中心（前Alberti医院）、位于阿尔巴尼亚的地拉那之塔（Tower of Tirana）、位于阿拉伯联合酋长国杜拜的KPM Tower和“Meravigliosa Island”项目。

在工业设计领域，事务所获得了照明和建筑材料产品和部件设计方面的数量众多的专利，包括在一家合资企业内与两家行业领先公司共同开发的重构型赤陶的设计工作。这些项目都涉及了与建筑材料行业的深入合作，在设计业与制造业之间实施了实际而有效的协作。这些工作给事务所提供了在运用材料的基础上累积经验和开展研究的机会。

阿克雅建筑设计咨询有限公司在 佛罗伦萨、罗马、米兰、北京、 迪拜、圣保罗都设有工作室，有超过80位员工在辛勤工作着，其中包括了建筑师、设计师和图形设计人员，还有秘书、行政管理人员以及公共关系经理，他们来自于全球各个大学和各个国家。

PROJECTS LIST

- Pino Lo Jacono Art Gallery, Empoli, (FI), Italy, 1988.
- Private home in Via Frusa, Florence, Italy, 1988.
- Silvana clothing store, Montespertoli, (FI), Italy, 1988.
- Villa in Bagno a Ripoli, (FI), Italy, 1988.
- Il Parione, bar, Florence, Italy, 1988.
- Design for a river park on the Arno, Florence, Italy, 1989.
- Private home Costa San Giorgio, Florence, Italy, 1990.
- Tornabuoni Arte Contemporary Art Gallery, Lungarno B. Cellini, Florence, Italy, 1991.
- New offices for Studio Archea, Lungarno B. Cellini, Florence, Italy, 1991.
- Design for a Residential and Artisan Center, Castiglioncello, Livorno, Italy, 1992.
- Design for a master plan for the S.D.O., Eastern Business and Administrative System of Rome, Italy, 1992.
- Stop Line, center for entertainment and dining, Curno, (BG), Italy, 1994-1995.
- Plan for area of Mons. Natale Basilico, square and parking area, Merate, (LC), Italy, 1996.
- New Municipal Library with connected auditorium, Curno, (BG), Italy, 1996.
- Santo Ficara art gallery, Florence, Italy, 1997.
- Single-family private home in Leffe, (BG), Italy, 1997.
- Design for Nuvolari discotheque, Cremona, Italy, 1997.
- Swimming Pool in Val Seriana, Bergamo, Italy, 1997.
- Pezzoli Shop, Bergamo, Italy, 1997.
- Shopping center, Nembro, (BG), Italy, 1998.
- Pier Capponi office center, Florence, Italy, 1998.
- Villa in Gazzaniga, (BG), Italy, 1998.
- Qlò discotheque in Fiesole, Florence, Italy, 1998.
- General design for a new discotheque in Bellaria Igea Marina, (RN), Italy, 1998.
- Urban improvement plan for Tavarnuzze, (FI), Italy, 1999.
- Expansion of port of Fontville, Montecarlo, 1999.
- Design for a commercial equipment center in San Maurizio D'Opaglio, (NO), Italy, 1999.
- Design for the "Museo del Gusto", food museum, Impruneta, (FI), Italy, 1999.
- Design for a residential division, Pozzolatico, (FI), Italy, 2000.
- New D'Avenza factory, Massa Carrara, Italy, 2000.
- Remodeling of Tornabuoni Arte contemporary art gallery, Portofino, (GE), Italy, 2000.
- Remodeling of a private single-family home in Lugano, Switzerland, 2000.
- Design for a multi-use cultural center in Merate, (LC), Italy, 2000.
- Design for renovation of via Tirreno, Potenza, Italy, 2001.
- Commercial and residential center, Tavarnuzze, (FI), Italy, 2001.
- Body's Gym, Florence, Italy, 2001.
- Design of the Martini Illuminazione S.p.A. stand at Euroluce, Milan, Italy, 2001.
- Remodeling and conversion of former Officine Lenzi industrial complex, Lucca, Italy, 2001.
- Executive design of Milan-Naples highway for projects in the Impruneta and Scandicci areas – Bottai parking, cycling-pedestrian paths, landscaping between the road and the Greve River, improvement of Parco Pali area, pedestrian overpass-Area Certosa, Florence, Italy, 2001.
- Executive design of Milan-Naples highway for projects in the Impruneta and Scandicci areas – transportation hub, Certosa area, Florence, Italy, 2001.
- Executive design of Milan-Naples highway – Restoration, landscaping and environmental inclusion, (FI), Italy, 2001.
- Design for installing sound-absorbing barriers on a section of the Milan-Naples highway, Italy, 2001.
- Alzheimer's Center, Foligno, (PG), Italy, 2001.
- Art park in Bagno a Ripoli, (FI), Italy, 2001.
- Municipal Library, Nembro, (BG), Italy, 2002.
- New GranitiFiandre showroom, Castellarano, (RE), Italy, 2002.
- New Radici Chimica office building, Linz, Austria, 2002.

项目列表

- 加科诺艺术画廊，恩波利（FI），意大利，1988年。
- 维亚弗鲁莎私人住宅，佛罗伦萨，意大利，1988年。
- 西尔瓦娜服装店，蒙特斯佩托（FI），意大利，1988年。
- Bagno a Ripoli的别墅，（FI），意大利，1988年。
- 伊尔帕里诺，酒吧，佛罗伦萨，意大利，1988年。
- 阿尔诺的河边公园设计，佛罗伦萨，意大利，1989年。
- 科斯塔圣乔治的私人住宅，佛罗伦萨，意大利，1990年。
- Tornabuoni当代艺术画廊，Lungarno B. Cellini，佛罗伦萨，意大利，1991年。
- 阿克雅工作室新办公室，Lungarno B. Cellini，佛罗伦萨，意大利，1991年。
- 住宅和工艺中心设计，Castiglioncello，里窝那，意大利，1992年。
- S.D.O.主要规划设计，罗马东方商务管理系统，意大利，1992。
- 停车线，娱乐和餐饮中心，Curno，（B.G.），意大利,1994 – 1995年。
- 蒙斯地区规划。纳塔尔巴西利科，广场和停车场，梅拉泰（LC），意大利，1996年。
- 与礼堂相连的新市政图书馆，Curno，（B.G.），意大利，1996年。
- 桑托菲卡拉美术馆，佛罗伦萨，意大利，1997年。
- Leffe的单个家庭的私宅，（B.G.），意大利，1997年。
- 诺瓦拉利迪斯科舞厅设计，克雷莫纳，意大利，1997年。
- 维尔塞里亚纳游泳池，贝加莫，意大利，1997年。
- 佩佐利商店，贝加莫，意大利，1997年。
- 购物中心，尼姆罗，（B.G.），意大利，1998年。
- 皮埃尔卡帕尼办公中心，佛罗伦萨，意大利，1998年。
- 加扎尼加的别墅（B.G.）,意大利,1998年。
- 费索罗Ql ò 迪斯科舞厅,佛罗伦萨,意大利,1998年。
- 贝拉里亚伊贾马里纳新迪斯科舞厅的整体设计,（RN）,意大利,1998年。
- Tavarnuzze的城市改善计划,（FI）,意大利,1999年。
- Fontville港的扩展计划,蒙特卡洛,1999年。
- 莫里吉奥德奥帕格里奥商业装备中心的设计（NO），意大利,1999年。
- “Museo del Gusto”的设计,食物博物馆, Impruneta,(FI), 意大利,1999年。
- 居住区设计,Pozzolatico,(FI), 意大利, 2000年。
- 新D’Avenza工厂,Massa Carrara,意大利,2000年。
- Tornabuoni Arte 现代艺术馆的改建, Portofino, (GE), 意大利, 2000年。
- 卢加诺独户私人住宅的改建,瑞士, 2000年。
- 梅拉泰多用途文化中心设计,意大利,2000年。
- 维亚泰拉诺的改建设计,意大利,2001年。
- 商业和居住中心,泰瓦努兹,意大利,2001年。
- 身体健身馆,佛罗伦萨,意大利,2001年。
- 尤拉卢斯Martini Illuminazione S.p.A. 的设计,米兰,意大利,2001年。
- 兰兹工业区原办公室的重建和改建,卢卡,意大利,2001年。
- Impruneta 和 Scandicci 区域米兰那不勒斯高速公路的执行设计,博塔公园,自行车赛路,道路与格里夫河之间的景观设计,Parco Pali 地区的改,Certosa区域的天桥,佛罗伦萨,2001年。
- Impruneta 和 Scandicci 区域米兰那不勒斯高速公路的执行设计,交通枢纽,色托沙地区,佛罗伦萨,意大利，2001年。
- 米兰–那不勒斯高速公路执行设计– 修复,景观和环境规划,(FI),意大利,2001年。
- 米兰–那不勒斯高速公路吸音屏障的安装设计,意大利,2001年。
- Alzheimer中心，菲拉哥诺, 意大利, 2001年。
- Bagno a Ripoli艺术公园,(FI),意大利,2001年。

- Programmatic plan for the new regional masterplan for Assisi, (PG), Italy, 2002.
- Design for the conversion of the former Fiat area in Novoli, Florence, Italy, 2002.
- Design, renovation and adaptation to regulations of the Campeggio Michelangelo campground, Florence, Italy, 2003.
- Entrance piece at the 50th International Art Exhibition, Venice Biennale, "The Cord", Venice, Italy, 2003.
- Cersaie stand 2003 - GranitiFiandre, Bologna, 2003.
- Design for the conversion of a convent into a library, Canicattì, (AG), Italy, 2003.
- Design of "Enzimi" cultural event, Rome, Italy, 2003.
- Design of the Martini Illuminazione SpA stand at Euroluce, Milan, Italy, 2003
- Interior design of pavilion for introducing the new Renault Megane, Rome, Italy, 2003.
- Design for the former Lazzeri theatre, Livorno, Italy, 2003.
- Urban renewal of Salaroli area, Cesena, Italy, 2003.
- Remodeling of Podere Africa, Siena, Italy, 2004.
- Schematic design for an Olympic village in Doha for the Asian Games 2006, Qatar, 2004.
- Design for a day center for disabled people and community housing, Seregno, (MI), Italy, 2004.
- Conversion of complex of former military hospital, Trieste, Italy, 2004.
- Conversion of former Metropolitan cinema into a shopping gallery and housing, Livorno, Italy, 2004.
- "Floating City" pavilion – Venice Bienniale, Italy, 2004.
- Restoration of Ex Panificio Militare area, Florence, Italy, 2004.
- Renewal project for Via Cesare Battisti, Cesena, Italy, 2004.
- Design for Villa le Fonti Hotel and Spa, Monteloro, Florence, Italy, 2004.
- New GranitiFiandre Meeting Room, Castellarano, (RE), Italy, 2005.
- Municipal urban plan for Finale Ligure, (SV), Italy, 2005.
- Proposal for reorganization and expansion of Perfetti Van Melle facility in Lainate, Milan, Italy, 2005.
- Design for a new division in Loc. Diacceto, Pelago, (FI), Italy, 2005.
- Design for the new Cantina Antinori winery in Bargino, San Casciano Val di Pesa, (FI), Italy, 2005.
- Multi-storey Parking, Piazzale Caduti della Polizia, Cesena, Italy, 2005.
- Underground parking behind Palazzo Tettamanti, Piazza degli Eroi, Merate, (LC), Italy, 2005.
- Masterplan concept proposal for a new city in Cheng Du, China, 2005.
- "Borgo Arnolfo", New residential and commercial center, former Alberti hospital, San Giovanni Valdarno, (FI), Italy, 2006.
- Design for exhibition-event "Vinar Vino, Arte e Architettura" Stazione Leopolda, Florence, Italy, 2006
- Commissioned to design renovation of Cornigliano, Genoa, Italy, 2006.
- Urban renovation of the former Ticosa, Como, Italy, 2006.
- Design for an auditorium, ex Magazzino vini Trieste, Italy, 2006.
- Design for new offices of the Italian Chamber of Commerce in Beijing, China, 2006.
- Design for the structural remodeling of the former Albergo Firenze Nova hotel, Florence, Italy, 2006.
- Masterplan of Castello, Florence, Italy, 2006.
- Remodeling and expansion of a villa, Bagno a Ripoli, (FI), Italy, 2006.
- New industrial area San Zeno, Arezzo, Italy, 2006.
- Design to build 26 units for private homes in Località Sa Carrubedda, Sant'Anna Arresi, (CI), Italy, 2006.
- Design proposal for a residential complex in Lainate,

• 市政图书馆, 纳姆博罗,(BG), 意大利,2002年。
• 新GranitiFiandre剧场,Castellarano,(RE),意大利,2002年。
• 新Radici Chimica办公室建筑,林兹,奥地利,2002年。
• 新阿西西区域总体规划，（PG），意大利，2002年。
• 诺沃利前菲亚特地区设计，佛罗伦萨，意大利，2002年。
• 坎佩焦米开朗基罗露营地的设计、改建和附建，佛罗伦萨，意大利，2003年。
• 第50届国际艺术展入口设计，威尼斯双年展，“线”，威尼斯，意大利，2003年。
• Cersaie站2003年 - GranitiFiandre，博洛尼亚，2003年。
• 将原建筑改建成一个图书馆，卡尼卡蒂，（AG），意大利，2003年。
• “Enzimi”文化活动设计，罗马，意大利，2003年。
• 欧罗卢斯马提尼Illuminazione水疗中心设计，米兰，意大利，2003年。
• 介绍新梅甘娜雷诺的室内设计展区，罗马，意大利，2003年。
• 前拉泽里剧院的设计，里窝那，意大利，2003年。
• 萨拉罗利领域市区重建，切塞纳，意大利，2003年。
• Podere住房重建，锡耶纳，意大利，2004年。
• 原理图设计，多哈2006年亚运会亚运村的设计，卡塔尔，2004年。
• 适合残疾人士和社会住房的设计，塞雷尼奥，（MI），意大利，2004年。
• 前军医院的改建，的里雅斯特，意大利，2004年。
• 前大都会电影院改建为购物商场及房屋，里窝那，意大利，2004年。
• “浮城”展馆，威尼斯双年展，意大利，2004年。
• 前军区面包厂房区的重建设计，佛罗伦萨，意大利，2004年。
• 维亚切萨雷巴蒂斯迪的重建项目，切塞纳，意大利，2004年。
• 乐丰蒂酒店及水疗中心别墅设计，Monteloro，佛罗伦萨，意大利，2004年。
• 新GranitiFiandre会议室，卡斯特拉腊诺，（RE），意大利，2005。
• 菲纳莱利古雷城市规划，（SV），意大利，2005年。
• 重组建议和Lainate不凡帝范梅勒扩建设计，意大利，米兰，2005年。
• 洛可迪亚塞托新区域的设计，（FI），意大利，2005年。
• Bargino的新坎蒂娜安蒂诺里酒厂设计，圣卡夏诺塔瓦内莱，（FI），意大利，2005年。
• 多层停车场，卡都提德拉普利西亚广场，切塞纳，意大利，2005年。
• 宫泰塔曼蒂后地下停车场，匹亚扎德格里伊亚罗，梅拉泰（LC），意大利，2005年。
• 成都新城的总体规划，成都，中国，2005年。
• “博尔戈阿诺尔福”新住宅及商业中心，前阿尔贝蒂医院，圣乔瓦尼达诺，（FI），意大利，2006年。
• 酿酒葡萄，艺术与建筑展览前Leopolda火车站，佛罗伦萨，意大利，2006年。
• 受委托设计Cornigliano，热那亚，意大利，2006年。
• 前Ticosa城市改造，科莫，意大利，2006年。
• 前藏酒仓库改造小礼堂设计，意大利，2006年。
• 意大利商会在北京新的办事处，中国，2006年。
• 前佛罗伦萨店重建，佛罗伦萨，意大利，2006年。
• Castello区总体规划，佛罗伦萨，意大利，2006年。
• 巴尼奥阿里利波里的别墅修复与扩建，（FI），意大利，2006年。
• 圣芝诺新的工业区，阿雷佐，意大利，2006年。
• 卢卡里塔私人住宅的26个单元设计。Sa Carrubedda，Sant'Anna Arresi，（CI），意大

Milan, Italy, 2006.
· Proposal for a service complex at the university, Marghera, Venice, Italy, 2006.
· Design for a new pedestrian and vehicle bridge in Imola, Italy, 2006.
· Design for 4 villas in Merate, (LC), Italy, 2006.
· Xujiahui Five Towers Project, Shanghai, China, 2006.
· SS100 Changsha Show Flat, Changhsha, China, 2006.
· SS100 Shenyang Show Flat, Shenyang, China, 2006.
· Houhai Italian Restaurant, Beijing, China, 2006.
· Archea Beijing New Office, Beijing, China, 2006.
· Design for Tango discotheque, Beijing, China, 2006-2007.
· New World entrance, Beijing, China, 2006.
· Façade restyling proposal, Inner Mongolia, China, 2006.
· New World office building, Beijing, China, 2006.
· Meishan residential masterplan, Meishan, China, 2006.
· New offices for Marco Polo company in Beijing, China, 2006-2007.
· SS100 Chengdu Landscape, Chengdu, China, 2006.
· Jiayuan Masterplan, Nanning, China, 2006.
· Huafa Interior Design, Beijing, China, 2006.
· Annali dell'Architettura e delle città, Palazzo Reale, Naples, Italy, 2006.
· Design for the new Berlucchi winery, Borgonovo di Corte Franca, (BS), Italy, 2006.
· Remodeling of property at Via S. Canterina da Siena, 10 Merate, (LC), Italy, 2006.
· Design for a riverside park by the Ciuffenna River in Terranuova Bracciolini, (AR), Italy, 2006.
· Design for the Campeggio Albatros campground, San Vincenzo, (LI), Italy, 2007.
· Design for an artisan development, Follonica, (GR), Italy, 2007.
· Bicu, pub-restaurant, Genoa, Italy, 2007.
· Design for a subsidized residential housing building and space for production of diverse types, in the area of a former municipal workshop, Sesto Fiorentino, (FI), Italy, 2007.
· Design for the "Torre delle Arti" (Tower of Arts), Milan, Italy, 2007.
· New urban layout for area of former freight depots of Valdellora and Acam, La Spezia, Italy, 2007.
· Feasibility plan for the renovation and expansion of the former rehabilitation center "Nicò", Pontedera, (PI), Italy, 2007.
· Residential, commercial and hospitality project in the former Fonderie Bigagli area, Prato, Italy, 2007.
· Architectural consulting for a new system of energy and agronomic optimization in Vignolo, (CN), Italy, 2007.
· Preliminary design for the Manifatture Tabacchi area for Pirelli SpA, Milan, Italy, 2007.
· Preliminary design for the Manifatture Sigaro Toscano area, Cava dei Tirreni, (SA), Italy, 2007.
· Masterplan Dianduchang, Xixia District, Nanjing, China 2007.
· SS100 Yantai Show Flat, Yantai, China, 2006-2007.
· P.U.G. - General (structural and programmatic) urban plan for Bisceglie, (BA), Italy, 2007.
· Design for the Expo in Shanghai 2010, Shanghai, China, 2007.
· BPT Control Room, Beijing, China, 2007.
· Bergamo Masterplan, Bergamo, Italy, 2007.
· Annali dell'Architettura e delle città, Palazzo Reale, Naples, Italy, 2007.
· Zhuzhou Commercial Buildings, Zhuzhou, China, 2007.
· Stoolcase Furniture Design, Beijing, China, 2007.
· FIAT new offices, Shanghai, China, 2007.
· Black Club, Tianjin, China, 2007.
· Design for the new Birra Peroni headquarters, Rome, Italy, 2007.
· Rizzoli office renovation, Beijing, China, 2007.
· SMIC Masterplan Project, Tianjin, China, 2007.

利，2006年。
▪ Lainate住宅区设计提案，米兰，意大利，2006年。
▪ 大学服务机构设计建议， Marghera，威尼斯，意大利，2006年。
▪ 伊莫拉人行、车辆桥的设计，意大利，2006年。
▪ 梅拉泰四栋别墅的设计，（LC），意大利，2006年。
▪ 徐家汇五塔项目，上海，中国，2006年。
▪ SS100长沙示范单位，长沙，中国，2006年。
▪ SS100沈阳示范单位，沈阳，中国，2006年。
▪ 后海的意大利餐厅，北京，中国，2006年。
▪ 阿克雅北京新办事处，北京，中国，2006。
▪ 探戈舞厅，北京，中国，2006-2007年。
▪ 新世界的入口设计，北京，中国，2006年
▪ 外墙改装的建议，内蒙古，中国，2006年。
▪ 新世界办公楼，北京，中国，2006年。
▪ 梅山住宅总设计，梅山，中国，2006年。
▪ 新马可波罗公司的办公室，北京，中国，2006-2007年。
▪ SS100成都景观，成都，中国，2006年。
▪ 佳苑总体规划，南宁，中国，2006年。
▪ 华发室内设计，北京，中国，2006年。
▪ 阿纳利德尔阿基泰克塔阿尔西塔的设计，帕拉佐里尔，那不勒斯，意大利，2006年。
▪ 新贝卢基酒厂设计，博尔戈诺沃迪科尔特弗兰卡（BS），意大利，2006年。
▪ 瑞纳达西纳路10号不动产的再设计，梅拉泰，（LC），意大利，2006年。
▪ 泰拉努瓦布拉乔利尼的奎菲纳河河滨公园，（AR），意大利，2006年。
▪ 坎佩焦信天翁露营地的设计，圣温琴，（LI），意大利，2007年。
▪ 技术工人发展的设计，福洛尼卡，（GR），意大利，2007年。
▪ 比库，酒吧，餐厅，热那亚，意大利，2007年。
▪ 原市政工厂居住空间和多功能生产空间设计，塞斯托菲伦蒂诺，佛罗伦萨（FI），意大利，2007年。
▪ “托瑞阿尔阿尔季”（艺术大楼）的设计，米兰，意大利，2007年。
▪ 瓦德罗那和阿查姆，拉斯佩齐亚的新城市布局，意大利，2007年。
▪ 前康复中心扩建计划可行性研究，尼科，蓬泰代拉，（PI），意大利，2007年。
▪ 前冯特利尔比加利区住宅、商业及酒店项目，普拉托，意大利，2007年。
▪ 维格诺洛新能源和农艺优化系统的设计顾问，（CN），意大利，2007年。
▪塔巴基的倍耐力SpA公司的初步设计，意大利米兰，2007年。
▪ 美尼法图西格罗托斯卡诺区初步设计，卡瓦德泰雷尼，（SA），意大利，2007年。
▪ 南京栖霞区电镀厂设计，南京，中国，2007年。
▪ SS100烟台示范单位，烟台，中国，2006-2007年。
▪ P.U.G. – 一般设计（结构和方案）
比斯塞格力城市规划，（BA），意大利，2007年。
▪ 2010年上海世博会设计，上海， 中国，2007。
▪ BPT的控制室，北京，中国，2007。
▪ 贝加莫总体规划，贝加莫，意大利，2007年。
▪建筑与城市年表，皇宫，那不勒斯，意大利，2007年。
▪ 株洲商业建筑物，株洲，中国，2007。
▪ Stoolcase家具设计，北京，中国，2007年。
▪ 菲亚特新的办事处，上海，中国，2007。
▪ 黑色俱乐部，天津，中国，2007年。
▪ 新贝洛尼啤酒总部的设计，意大利罗马，2007年。
▪ 里佐利办公室装修，北京，中国，2007年。
▪ 中芯国际总体规划项目，天津，中国，2007年。
▪ “跑酷 – 北京”住宅及商业中心，北京，中国，2007年。
▪ 飞鸭新办事处，北京，中国，2007。
▪ Gluhovo住宅项目，Gluhovo，俄罗斯，2007年。
▪ 交通大学GEL大厦，上海，中国，2007年。

- "Parkour - Beijing" residential and commercial center, Beijing, China 2007.
- PIAP New Office, Beijing, China, 2007.
- Gluhovo Residential Project, Gluhovo, Russia, 2007.
- Jiaotong University GEL Building, Shanghai, China, 2007.
- Romana Ice Cream Shop, Beijing, China, 2007.
- Tethis New Office, Beijing, China, 2007.
- Design for 10 lots on Via dei Barberi - Zona PEEP, Planning for the Landscape, Environment and Infrastructure, Grosseto, Italy, 2007.
- Landscaping plan for design "I Cipressi" waste-to-energy plant in the municipality of Rufina, Località Selvapiana, (FI), 2007.
- Milanfiori 2000, design ideas for a residential complex, Assago, (MI), Italy, 2007.
- Conversion plan for the former municipal workshop area, Sesto Fiorentino, (FI), 2006.
- Design for the new municipal offices and library of Figline Val d'Arno, (FI), Italy, 2007.
- Design of building for subsidized housing, Sesto Fiorentino, (FI), Italy, 2007.
- Design of exhibition "Learning from South", Arsenale Fortezza da Basso, Florence, Italy, 2007.
- Design of the Martini Illuminazione SpA stand at Euroluce, Milan, Italy, 2007
- Design for the exhibition "The House of the III Millennium", Palazzo Novellucci, Prato, Italy, 2007.
- Expansion of Villa Il Casale, Greve in Chianti, (FI), Italy, 2008.
- Annali dell'Architettura e delle Città, Palazzo Reale, Naples, Italy, 2008.
- Design for the exhibition "The House of the 3rd Millennium", Palazzo Novellucci Urban Center, Prato, Italy, 2008.
- Lagoon Park_She[l]ter, event at the 11th Venice Biennale International Architecture Exhibition, Venice, Italy, 2008.
- Design for a tower, Modena, Italy, 2008.
- State Police Multi-use Center, Naples, Italy, 2008.
- Martini showroom, Foro Bonaparte, Milan, Italy, 2008.
- Tirana Boulevard Cavages Albania, 2008.
- Design for a residence in Dubai, United Arab Emirates, 2008.
- New Masterplan for Tripoli, Libya, 2008.
- Design for the Forte Carlo Felice Residence, La Maddalena, Olbia Tempio, (SS), Italy, 2008.
- Executive urban plan in Padua, Italy, 2008.
- Design for the KPM Tower, Dubai, United Arab Emirates, 2009.
- Art Cube, Casalbeltrame, Novara, Italy, 2009.
- New University of Tripoli, Libya, 2009.
- Design for Meravigliosa Island, "The World", Dubai, United Arab Emirates, 2009.
- Design for Villa Lusini – Portoferraio Isola D'Elba, (LI), Italy, 2009.
- Design for former Manifattura Sigaro Toscano, Foiano della Chiana, (AR), Italy, 2009.
- Design for the Pub Bicu, Savignano sul Rubicone, (FC), Italy, 2009.
- Design for the UBPA B3 Pavilion, Expo Shanghai, China, 2010
- Design for the Chang-Li Winery, China, 2010
- Design for the Nanping Village, China, 2010
- Design for the residential complex Colel Loreto, Lugano, Switzerland, 2011
- Design for an hotel in Doha, Qatar, 2011
- Design for Villa Verde complex, São Paulo, Brasil, 2011
- Design for the Li Ling Ceramic Hotel, China, 2011
- Design for the Tasly Hotel, Tyanjin, China, 2011

- 罗马纳冰淇淋店，北京，中国，2007年。
- Tethis新办事处，北京，中国，2007年。
- 维亚德巴伯里10年地段的设计，祖纳皮帕，景观、环境和基础设施设计，格罗塞托，意大利，2007年。
- 园林绿化设计方案“，“艾克拉皮西”，卢菲纳废物转化能源工厂，卢科里塔塞尔瓦皮亚那，（FI），2007年。
- Milanfiori 2000，住宅区设计理念，阿萨哥，（MI），意大利，2007年。
- 原市场工厂区域转换计划，塞斯托菲尔纳托，佛罗伦萨，（FI），2006年。
- 新的市政办公室和菲利内瓦尔瓦尔德阿诺图书馆，（FI），意大利，2007年。
- 福利住宅的设计，塞斯托菲尔纳托，佛罗伦萨，（FI），意大利，2007年。
- “向南部学习”展览设计，阿森纳尔弗特扎达巴索，佛罗伦萨，意大利，2007年。
- 尤拉卢斯的马提尼Illuminazione水疗中心设计，米兰，意大利，2007年。
- “三千年住宅”展览设计，诺委罗齐宫，普拉托，意大利，2007年。
- 卡塞尔二期别墅扩建，奇安提格雷福，（FI），意大利，2008年。
- 建筑与城市年表，皇宫，那不勒斯，意大利，2008年。
- “三千年住宅”展览设计诺委罗齐宫，普拉托，意大利，2007年。
- 第11届国际建筑威尼斯双年展，拉贡公园，威尼斯，意大利，2008年。
- 一座高层的设计，摩德纳，意大利，2008年。
- 国家警察多用途中心，那不勒斯，意大利，2008年。
- 马提尼陈列室，福罗波拿巴，米兰，意大利，2008年。
- 阿尔巴尼亚地拉那大道，卡瓦吉斯， 2008年。
- 在迪拜设计住宅，阿拉伯联合酋长国，2008年。
- 的黎波里新的总体规划，利比亚，2008年。
- 卡罗菲利斯堡别墅的设计，香格里拉马达列纳，奥尔比亚滕皮奥，（SS），意大利，2008年。
- 帕多瓦城市规划执行设计，意大利，2008年。
- PKM塔设计，迪拜，阿拉伯联合酋长国，2009年。
- 艺术立方，卡萨白达玛，诺瓦拉，意大利，2009年。
- 的黎波里新大学，利比亚，2009年。
- Meravigliosa岛的设计，“世界”，迪拜，阿拉伯联合酋长国，2009年。
- 卢西尼别墅设计， 费拉约港伊索拉德埃尔巴，（LI），意大利，2009年。
- 前马尼法图拉西哥尼托斯卡诺设计，西亚诺德拉奇亚那，（AR），意大利，2009年。
- 比库酒吧的设计，萨维尼亚诺萨维尼亚诺，（FC），意大利，2009年。
- 城市最佳实践区,上海世界博览会,中国，2010年 。
- 昌黎酒庄设计, 中国,2010 年 。
- 南平村设计,中国,2010年 。
- 居住群设计, 卢加诺, 瑞士,2011年 。
- 酒店设计, 多哈,卡塔尔, 2011年 。
- 别墅群设计, 圣保罗, 巴西, 2011年 。
- 醴陵陶瓷酒店, 中国, 2011年 。

COMPETITIONS

- "Forum" Tokyo ideas competition, China, 1989.
- "Tre chiese per il 2000" ideas competition, 1989.
- Ideas competition for the "Santa Verdiana new university cafeteria" in Florence, Italy, 1989.
- Ideas competition for the new "Acropolis museum", Athens, 1989.
- International ideas competition for the conversion of the former Fossoli concentration camp in Carpi, (MO), Italy, 1989, jury mention.
- Venice Biennale international competition "Una porta per Venezia", Italy, 1991.
- International ideas competition for the urban plan of the Spreebogen area, Berlin, 1992.
- Todi competition, (PG), Italy, 1992.
- Competition for a housing model, on behalf of the IACP of the province of Treviso, a residential district in Castelfranco Veneto, (TV), Italy, 1992.
- Ideas competition for the "Lungocanale di Cesenatico", (FC), Italy, 1992, with Loris Macci.
- National architecture competition "Una via, tre piazze", Gela, (CL), Italy, 1993.
- Competition for building an educational center in Piedicastello, (TN), Italy, 1993.
- Europan competition, Quarrata project area, (PT), Italy, 1993.
- Competition to design two parish centers for the Dioceses of Rome as part of the "50 chiese for the duemila" (50 churches for 2000), program, Italy, 1994.
- Ideas competition for a new block in Gutembergstraße, Kiel, Germany, 1996, with Loris Macci.
- Ideas competition "Polo Espositivo Unitario e Integrato di Brescia" exhibition center, Italy, 1996, 4th place, jury mention.
- Competition for a new plan for the "Piazza della Vittoria" in Castiglioncello, (LI), Italy, 1993, 1st place design.
- Barcelona Museum competition, Spain, 1999.
- International competition Kansai-kan of the National Diet Library, Japan, 1996, honorable mention.
- Competition for the preliminary design o the ZIPA administrative center of Iesi, (AN), Italy, 1997, 1st place.
- Competition for building a new urban gateway for Parma, Italy, 1997, with Aurelio Cortesi, 2nd place.
- Competition for the new ATAF cafeteria, Florence, Italy, 1997, 1st place.
- Competition for designing the new Contemporary Art Museum, area ex-Fergat in Turin, Italy, 1998
- "Luigi Cosenza National Prize" Competition, Naples, Italy, 1998, honorable mention.
- Competition for a new IUAV center in the Frigoriferi warehouse area in San Basilio, Venice, Italy, 1998, 3rd place.
- National competition for an administrative and commercial center in Calenzano, (FI), Italy, 1998, with Ipostudio and Elio di Franco, 1st place.
- International architecture competition "La città della scuola a Sarno" (School city in Sarno), (SA), Italy, 1999.
- Competition for the new entrance to the civic museum and definition of surrounding area, municipality of Padua, Italy, 1999.
- Competition for the "New headquarters of the Agenzia Spaziale Italiana" (Italian space agency) in Rome, Italy, 2000, with Franz Prati, 2nd place.
- European competition for the design of a new cultural center in Cinisello Balsamo, Milan, Italy, 2001.
- Competition for the expansion of the museum and square in Santa Corona, Vicenza, Italy, 2001.
- Competition for the design of the Chiesa del Redentore in Nonantola, Modena, Italy, 2001, 3rd place.
- Competition for the new travel center for the Porta Susa station in Turin, Italy, 2001, among the seven designs selected for the second phase
- Competition to convert the former area Fiat in Viale

赛事

- “论坛”东京创意竞赛，中国，1989年。
- “2000年的三个教堂”创意竞赛，1989年。
- “圣维地亚纳新大学食堂”设计理念竞赛，佛罗伦萨，意大利，1989年。
- “雅典卫城博物馆” 新思路竞赛，雅典，1989年。
- 前弗索罗集中营功能转换创意竞赛，卡皮，（MO），意大利， 1989年，评委会奖。
- 威尼斯双年展国际竞赛“给威尼斯的一扇门”， 威尼斯，意大利，1991年。
- Spreebogen区城市规划国际创意竞赛，柏林，1992年。
- Todi竞赛，（PG），意大利，1992年。
- 房屋模型竞赛，代表特维索省的IACP，卡斯泰尔弗朗科维聂托的住宅区，（TV），意大利，1992年。
- “Lungocanale di Cesenatico”创意竞赛, (FC), 意大利, 1992年, 与劳力斯・马西合作。
- “一条道路，三个广场”国家建筑设计竞赛, 哥拉, (CL), 意大利, 1993年。
- Piedicastello教育中心建设竞赛， (TN), 意大利, 1993年。
- 欧罗班竞赛，Quarrata项目区域, (PT), 意大利, 1993年。
- 为罗马的迪奥塞斯两个教区中心建设与进行的竞赛，属2000年50座教堂项目的一部分，规划，意大利，1994年。
- Gutembergstraße新街区建设竞赛，凯尔，德国，1996年, 与劳力斯・马西合作。
- “Polo Espositivo Unitario e Integrato di Brescia”展览中心创意竞赛， 1996年，第4名，评委会奖。
- 卡提格罗西奥“维多利亚广场”新规划竞赛，（LI），意大利，1993年，设计第一名。
- 巴塞罗那 博物馆竞赛，西班牙，1999年。
- 国际饮食图书馆Kansai-kan国际竞赛，日本，1996年，荣誉奖。
- 列西ZIPA行政中心初步设计竞赛，（AN），意大利，1997,第一名。
- 帕尔马新城市入口建筑的竞赛，意大利，1997年，与奥瑞罗・科泰西合作，第二名。
- ATAF新食堂设计竞赛，佛罗伦萨，意大利，1997年，第一名。
- 新现代艺术馆设计竞赛，图瑞恩ex-Fergat 地区，意大利，1998年。
- “Luigi Cosenza National Prize” 竞赛，那不勒斯，意大利，1998年，荣誉奖。
- 圣巴萨雷奥Frigoriferi 仓库新IUAV中心设计竞赛，威尼斯，意大利，1998年，第三名。
- 凯伦扎诺行政与商务中心国家竞赛，（FI），意大利，1998年，与艾波斯图地奥和阿里欧德弗朗科合作，第一名。
- “萨拉诺学校城” 国际设计大塞（萨诺大学城），（SA），意大利，1999年。
- 市民博物馆新入口设计竞赛以及周边区域定位，帕杜阿市，意大利，1999年。
- “意大利航空署新总部”设计竞赛（意大利航空署），罗马，意大利，2000年，与弗朗兹・普拉提合作，第二名。
- Cinisello Balsamo新文化中心欧洲设计竞赛，米兰，意大利，2001年。
- Santa Corona 博物馆和广场扩建竞赛，维琴察，意大利，2001年。
- Nonantola 的让帝多教堂设计竞赛，摩德纳，意大利，2001年，第三名。
- 图瑞尼新旅行中心Porta Susa站建设设计竞赛，意大利，2001年，进入第二轮的7项设计之一。
- 前菲亚特厂区转型设计竞赛，维勒。贝尔福勒，佛罗伦萨，意大利，2002年，荣誉奖。
- 米开朗基罗营地城市公园设计竞赛，佛罗伦萨，意大利，与莫罗・萨托合作，第一名。
- 佛罗伦萨新地铁设计竞赛，意大利，2002年。
- 高铁设计竞赛，“佛罗伦萨TAV站”，意大利，2002年，与Gruppo Toscano – Natalini Architetti、Elio Di Franco、Ipostudio、Claudio Nardi合作。

Belfiore, Florence, Italy, 2002, honorable mention.
· Competition for the City Park of Campeggio Michelangelo campground, Florence, Italy, with Mauro Saito, 1st place.
· Competition for the new underground station in Florence, Italy, 2002.
· High speed rail competition, "Florence T.A.V. station", Italy, 2002, with Gruppo Toscano - Natalini Architetti, Elio Di Franco, Ipostudio, Claudio Nardi.
· Competition for a "Multi-purpose public space", Savona, Italy, 2003, 1st place, with 5+1 AA.
· Competition for converting the "S. Colombano rampart", Lucca, Italy, 2003, 1st place.
· Competition for the "New Lazzeri theatre", Livorno, Italy, 2003, 1st place.
· Competition for "Rehabilitation of Piazza di Impruneta", (FI), Italy, 2003.
· International design competition for the "Centro europeo per le Creatività Emergenti", Pontecagnano Faiano, (SA), Italy, 2003, 1st place.
· Design competition the "Eugenio Battisti" museum of industry and labor, Brescia, Italy, 2004.
· International ideas competition, "La piazza della Basilica di Aquileia, rehabilitation of Piazza Capitolo and Piazza del Patriarcato", (UD), Italy, 2004, 3rd place.
· "Abitare a Milano", new urban places for social housing developments, Via Ovada, Milan, Italy, 2005, 3rd place.
· "Abitare a Milano", new urban places for social housing developments, Via Gallarate, Milan, Italy, 2005, 4th place
· Competition for the new "Palazzo del Cinema and surroundings" with architect Rafael Moneo, Venice, Italy, 2005.
· International design competition for "the former Area Ansaldo", Milan, Italy, 2005, tied for 1st place.
· Conversion of former cinema-theatre area and surroundings, Chiari, (BS), Italy, 2005, first and second phase.
· "Five Towers for Shanghai" project, Tower n. 2, Shanghai, China, 2005.
· International competition by inviation to build one of the ten towers planned in the masterplan for the new center of Tirana, Albania, 2005, 1st place.
· International design competition "menoèpiù2" - Divino Amore, daycare center, nursery school, square with bus terminal, public parking, Rome, Italy, 2005.
· Ideas competition for "rehabilitation of Cornigliano", (GE), Italy, 2005 1st place.
· Project to improve the landscape and nature along the Salerno - Reggio Calabria highway, including a highway museum, Italy, 2005, with Pietro Carlo Pellegrini and Studio Franchi Lunardini Partners, 1st place.
· Competition: "La Casa dell'Architetto", new site of the Order of Architects of Florence, Italy, 2005.
· Competition for the Waterfront of Latina, Italy, 2006.
· Competition for the a documentation and study center in Seveso, Italy, 2006, 3rd place.
· Competition by invitation to design new hotel facilities at the new exhibition center of the Milan Fair, Italy, 2006.
· Design competition for an elementary and nursery school in Bagno a Ripoli, (FI), Italy, 2006.
· International design competition "MOdaM" – museum and fashion school, Italy, 2006, first and second phases.
· Competition for the new Legal Center in Trento, Italy, 2006, first and second phases.
· Competition for the "new regional museum of Nuragic and contemporary art of the Mediterranean in Cagliari", Italy, 2006, with Franz Prati and MDU Architetti, mentioned by the jury.
· International design competition for an "Auditorium in Padua", 2006.
· Design competition for renovating and expanding the fair center in Baltera, Riva del Garda, (TN), Italy, 2006.
· Yanxi Commercial Center, Beijing, China, Italy, 2006

• 多功能公共空间设计竞赛，萨瓦纳，意大利，2003年，第一名，获5+1 AA奖。
• “S. Colombano rampart”功能转换设计竞赛，卢卡，意大利，2003年，第一名。
• “新Lazzeri剧场”设计竞赛，利沃诺，意大利，2003年，第一名。
• “mpruneta广场重建”设计竞赛，（FI），意大利，2003年。
• “欧洲新兴创造业中心”国际设计竞赛，波特卡那诺 法那诺，(SA)，意大利，2003年，第一名。
• “Eugenio Battisti”工业和劳动博物馆设计竞赛，布里斯加，意大利，2004年。
• “安奎拉大教堂广场，Capitolo广场与 Patriarcato广场重建”国际创意竞赛，(UD)，意大利，2004年，第三名。
• “住在米兰”竞赛,用于扩展社会居住空间，维尔奥瓦达，2005年，第三名。
• “住在米兰”竞赛,用于扩展社会居住空间，维亚加拉赖特，意大利，2005年，第四名。
• 与Rafael Moneo 合作参加“影院大楼和周边环境”设计竞赛，威尼斯，意大利，2005年。
• “前安萨尔多地区”国际设计竞赛，米兰，意大利，2005年，第一名。
• 前影剧院地区及周边环境转换设计，凯拉利，（BS），意大利，2005年，第一轮和第二轮。
• “上海五塔”工程，第N2塔，上海，中国，2005年。
• 泰拉纳新中心十塔之一的国际设计竞赛，阿尔巴尼亚，2005年，第一名。
• “menoèpiù2” 幼儿园、幼教学校、巴士终点广场、公共停车场国际设计竞赛，罗马，意大利，2005年。
• “Cornigliano”重建创意竞赛，（GE），意大利，2005年，第一名。
• S 高速公路沿线及高速公路博物馆景观发送设计竞赛，意大利，2005年，与Pietro Carlo Pellegrini和Franchi Lunardini工作室合作，第一名。
• “建筑师的家”设计竞赛，佛罗伦萨建筑师场馆新址，意大利，2005年。
• 拉提娜沃特弗朗德设计竞赛，意大利，2006年。
• 萨维索文件和研究中心设计竞赛，意大利，2006年，第三名。
• 米兰市场新展览中心新酒店设施设计邀请赛，意大利，2006年。
• Bagno a Ripoli 的小学、幼儿园设计竞赛，（FI）意大利，2006年。
• “MOdaM”国际设计竞赛，博物馆和时尚学校，意大利，2006年，第一轮和第二轮。
• 泰伦托新法律中心设计竞赛，意大利，2006年，第一轮和第二轮。
• “纳鲁加克新地区博物馆和卡格里亚里地中海现代艺术馆”设计竞赛，意大利，2006年，与Franz Prati和MDU Architetti合作，评委会奖。
• “帕杜瓦礼堂”国际设计竞赛，2006年。
• 巴尔提拉市场中心改、扩建设计竞赛，Riva del Garda, (TN), 意大利, 2006年。
• 延西商务中心，北京，中国，意大利，2006年。
• 高雄表演艺术中心，高雄，台湾，2006年。
• 湖畔建筑设计私人咨询，拉维诺，(VA), 意大利, 2006年。
• 玉林教育机构校园建设邀请赛，玉林，中国，2006年。
• 新运动城设计竞赛，陕西玉林，中国，2005年，第一名。
• 巴黎塔尔白克新法院设计国际创意大赛，Halle Freyssinet, ZAC Paris Rive Gauche, 法国, 2006年。
• 天津7号办公大厦设计竞赛，2006年，天津，第二名。
• 梅斯特历史中心地区改建国际设计竞赛，Compendio Umberto I°, 梅斯特，威尼斯，意大利，2006年。
• 武汉艺术博物馆，武汉，中国，2007年，第三名。

· Kaohsiung Performing Arts Center, Kaohsiung-Taiwan, 2006.
· Private consulting for architectural design of buildings on the lake front, Laveno, Area Ex Ceramica Lago, (VA), Italy, 2006.
· Competition by invitation for Yulin Education Campus, Yulin, China, 2006.
· Competition for new sports city, Shanxi Yulin, China, 2005, 1st place.
· International ideas competition for the new courthouse in Paris on the site of Tolbiac, Halle Freyssinet, ZAC Paris Rive Gauche, France, 2006.
· Competition "Tianjin Site No. 7 Office Buildings", 2006, Tianjin, 2nd place.
· International design competition for the renovation plan for an area including Mestre's historic center, Compendio Umberto I°, Mestre, Venice, Italy, 2006.
· Wuhan Art Museum, Wuhan, China, 2007, 3rd place.
· Ronchang Residential Competition, Changchun, China, 2007.
· Building U13 of the Milanofiori complex, Milan, Italy, 2007.
· Xisibei Preservation Strategy Project, Beijing, China, 2007.
· Tangshan Earthquake Memorial, Tangshan, China, 2007.
· Central Waterfront of Hong Kong, Cina, 2007.
· "East Shore Masterplan", Hothot, China, 2007, 4th place.
· "UBPA Pavilion Expo 2010", Shanghai, China, 2007, 2nd place.
· "Shangri-la Winery", Penglai, China, 2007.
· International design competition "Corso Como", Milan, Italy, 2007.
· Competition for the final design and execution of projects for the "Music and Culture Park", Porta al Prato, Florence, Italy, 2007, 2nd place.
· Competition "Shanxi Yulin Sport Center" 2006, 1st place.
· International design competition for a research tower in the industrial area of Padua, Italy, 2008.
· International competition for Cheogna City Tower, Korea, 2008.
· International competition for the design of a "synagogue in Potsdam", Germany 2008, special prize.
· International competition for the design of the "Chopin Museum", Warsaw, Poland, 2008.
· Competition for a "nursery school", Sanluri, (CA), Italy, 2008.
· Competition for a "Center of documentation on Nazism", Munich, Germany, 2009. Design chosen for second phase
· Competition for "Urban Center", Olbia, (SS), Italy, 2009.
· Competition for an "Agricultural center, Lecce, Italy, 2009.
· Changsha Mei xi lake international culture and art center Competition, 2011.
· Competition for the Guizhou Zhen Winery, Guizhou, China, 2011, 2nd place.

·荣昌居住设计大赛，长春，中国，2007年。
·Milanofiori U13号建筑设计，米兰，意大利，2007年。
·西斯北仓储战略项目，北京，中国，2007年。
·唐山地震纪念碑，唐山，中国，2007年。
·香港中心水岸，中国，2007年。
·“东岸总体规划”，呼合浩特，中国，2007年，第四名。
·“UBPA展馆2010世博会”上海，中国，2007年，第二名。
·“香格里拉葡萄酒厂”蓬莱，中国，2007年。
·“科莫路”国际设计大赛，米兰，意大利，2007年。
·“音乐、文化广场”最终设计和执行大赛，佛罗伦萨，意大利，2007年，第2名。
·“陕西玉林体育中心”设计大赛，2006年，第一名。
·帕杜瓦地区研究塔国际设计大赛，意大利，2008年。
·图格纳城塔国际设计大赛，韩国，2008年。
·“synagogue in Potsdam” 国际设计大赛，德国，2008年，特别奖。
·“肖邦纪念馆”设计国际大赛，华沙，波兰，2008年。
·“幼儿园”设计大赛， Sanluri, (CA), 意大利， 2008年。
·“纳粹文件中心”设计大赛，慕尼黑，德国，2009年，进入第二轮。
· “城市中心”设计大赛，奥尔比亚，（SS）意大利，2009年。
· “农业中心”设计大赛，莱切，意大利，2009年。
· 长沙梅溪湖文化艺术中心国际竞赛， 2011年。
· 贵州酒庄设计竞赛, 贵州，中国， 2011年，二等奖 。

EXHIBITIONS

- “La città Europea” traveling exhibition, 1991.
- “Habitat & Identità” exhibition/workshop, Arezzo, Italy, 1992.
- “Ambizioni e Visioni” Tuscan architects under 40 exhibition, 1992.
- Venice Biennale International Architecture Exhibition with Aurelio Cortesi, “La modernità nello spazio sacro”, Venezia, Italy, 1992-1993.
- “Dedicato a Roma” exhibition, Never-before-seen architecture exhibited at the Bocca della Verità, Rome, Italy, 1993.
- 6th Venice Biennale International Architecture Exhibition, Venice, Italy, 1996.
- “New Italian Architecture” traveling exhibition, Berlin, Germany, 1996.
- “Studio Archea” exhibition at the AAM Gallery in Milan, Italy. 1997.
- “New generation of Florentine architects” exhibition, Matera, Italy, 1997.
- “New Italian Architecture”, Buenos Aires, Brazil, 1998.
- “Venice and New Architecture” exhibition, Venice, Italy, 1999.
- 8th Venice Biennale International Architecture Exhibition, Venice, Italy, 2002.
- “From Futurism to a Possible Future” exhibition of Italian architecture, Japan, 2002.
- Traveling architecture exhibition in the cities of Campobasso, Benevento and Avellino, 2003.
- 9th Venice Biennale International Architecture Exhibition, “Floating City” pavilion, 2004.
- “Conflitti” exhibition, Salerno, 2005.
- “Laboratorio Italia” exhibition, profession and research section, Festival dell’Architettura, Parma, 2005.
- “13 muri di cartone per 13 architetti italiani” (13 cardboard walls for 13 Italian architects), exhibition part of the event “Città sottili, luoghi e progetti di cartone” (Thin cities, places and cardboard projects), curated by Pietro Carlo Pellegrini, Lucca, Italy, 2005.
- “Vinar. Vino, Arte e Architettura”, Stazione Leopolda, Florence, Italy, 2005.
- “Laboratorio Italia Roma 2006”, Aid’A Agenzia Italiana d’Architettura, DARC Direzione generale per l’architettura e l’arte contemporanee, Rome, Italy, 2006.
- “Annali dell’architettura e delle Città”, Naples, Italy, 2006.
- SS100 Quest of Happiness. Tianjin, China, 2006.
- Spazio FMG-Torre di Milano, Milan, Italy, 2007.
- “Annali dell’architettura e delle Città”, Naples, Italy, 2007.
- “House of the III Millennium”, Prato, Italy, 2007.
- Stoolcase Exhibition CIGE 2007, Beijing-China, 2007.
- Xisibei Regeneration Strategy Exhibition, Beijing, China, 2007.
- “Annali dell’architettura e delle Città”, Naples, Italy, 2008.
- “1968-2008, 40 years of design: continuity and discontinuity”, Milan, Italy, 2008
- “Brit Insurance Designs”, exhibition organized by the Design Museum, London, United Kingdom, 2008.
- “Ecological City/Building”, Beijing Biennial, China, 2008.
- “Sustainable Urbanization”, Libya, Spatial Development and Sustainability, Tripoli, Libya, 2009.
- “Dreaming Milano”, Milan, Italy, 2009.
- “L’invention de la tour européenne”, Pavillon de l’Arsenal, Paris, France, 2009.
- “AILATI Riflessi dal futuro” at the 12th Venice Biennale International Architecture Exhibition, Venice, Italy, 2010.
- “Sustainable Landmarks”, MuBe, Contemporary Brazilian Sculpture Museum of São Paulo, Brazil, 2011.
- “Italy Sustainable life”, Forte de Copacabana, Rio de Janeiro, Brazil, 2011.

展览

- “La città Europea”巡回展 , 1991年.
- “Habitat & Identità”展览/研讨会, Arezzo, 意大利, 1992年.
- “Ambizioni e Visioni” 40岁以下托斯卡建筑师展览, 1992年.
- 威尼斯双年展国际建筑艺术展览与Aurelio Cortesi, “La modernità nello spazio sacro”, Venezia, 意大利, 1992年-1993年.
- “Dedicato a Roma”展览, 在Bocca della Verità展出的前所未见的建筑艺术, 罗马, 意大利, 1993年.
- 第6届威尼斯双年展国际建筑艺术展览, 威尼斯, 意大利, 1996年.
- “新意大利建筑艺术”巡回展, 柏林, 德国, 1996年.
- “Studio Archea”展览, 在米兰AAM画廊举办, 意大利. 1997年.
- “新一代的佛罗伦萨建筑艺术”展览, 马特拉, 意大利, 1997年.
- “新意大利建筑艺术”, 布宜诺斯艾利斯, 巴西, 1998年.
- “威尼斯和新建筑艺术”展览, 威尼斯, 意大利, 1999年.
- 第8届威尼斯双年展国际建筑艺术展览, 威尼斯, 意大利, 2002年.
- “从未来主义到可能的未来”意大利建筑艺术展, 日本, 2002年.
- 建筑艺术巡回展, 在Campobasso、Benevento和Avellino市举办, 2003年.
- 第9届威尼斯双年展国际建筑艺术展览, “漂浮城市”展馆, 2004年.
- “Conflitti”展览, 萨勒诺, 2005年.
- “Laboratorio Italia”展览, 专业和研究部分, Festival dell’Architettura, 帕尔马, 2005年.
- “13 muri di cartone per 13 architetti italiani” (13位意大利建筑师的13座纸板墙), “Città sottili, luoghi e progetti di cartone” (瘦城市、场所和纸板项目)活动的展览部分, 由Pietro Carlo Pellegrini发起, 卢卡, 意大利, 2005年.
- “Vinar. Vino, Arte e Architettura”, Stazione Leopolda, 佛罗伦萨, 意大利, 2005年.
- “Laboratorio Italia Roma 2006”, Aid’A Agenzia Italiana d’Architettura, DARC Direzione generale per l’architettura e l’arte contemporanee, 罗马, 意大利, 2006年.
- “Annali dell’architettura e delle Città”, 那不勒斯, 意大利, 2006年.
- 阳光100寻访幸福人家. 天津, 中国, 2006年.
- Spazio FMG-Torre di Milano, 米兰, 意大利, 2007年.
- “Annali dell’architettura e delle Città”, 那不勒斯, 意大利, 2007年.
- “第三千年的房屋”, 普拉托, 意大利, 2007年.
- 2007年中艺博国际画廊展览会（CIGE 2007年）Stoolcase展, 中国北京, 2007年.
- 西四北改造战略展览会, 北京, 中国, 2007年.
- “Annali dell’architettura e delle Città”, 那不勒斯, 意大利, 2008年.
- “1968年-2008年, 设计40年:连续和间断”, 米兰, 意大利, 2008年
- “Brit Insurance Designs”, 由设计博物馆组织的展览, 伦敦, 英国, 2008.
- “生态城市/建筑”, 北京双年展, 中国, 2008年.
- “可持续城市化”, 利比亚, 空间开发和可持续性, 的黎波里, 利比亚, 2009年.
- “梦到米兰（Dreaming Milano）”, 米兰, 意大利, 2009年.
- “L’invention de la tour européenne”, Pavillon de l’Arsenal, 巴黎, 法国, 2009年.
- “AILATI Riflessi dal futuro” 威尼斯建筑双年展, 威尼斯, 意大利, 2010年
- “ 可持续地标 ”, Mube, 圣保罗巴西当代雕塑博物馆, 巴西, 2011 年

" 意大利可持续生活 ", Forte de Copacabana, 里约热内卢, 巴西, 2011 年

BIBLIOGRAPHY

· *Un'idea per le Murate, di Firenze*, Catalogue edited by Sergio Conti, Electa, Florence, 1989.
· Luigi Moiraghi, *Due progetti per il Forum di Tokyo*, in "Arca", no. 51, July-August 1991, pp. 28-37.
· Gianni Pettena, *In un interno fiorentino*, in "Domus", no. 730, September 1991, pp. 4-5.
· *Una porta per Venezia*, in "Materia", no. 8, III quarterly 1991, pp. 54-57.
· *L'idea: simmetria e classicità*, in "Interni", no. 416, September 1991, p. 160.
· Antonello Boschi, *Nuovo museo dell'Acropoli di Atene, Ad Atene piccola!*, in "P.A. Professione Architetto", no. 1, January 1992, pp. 40-47.
· *Una porta per Venezia*, in "Phalaris", no. 18, Anno IV, January-February 1992, pp. 29.
· Marco Casamonti, *Un simbolo antico per nuovo denaro, Recupero di un antico borgo a Todi*, in "Recuperare", no. 7, September 1992, pp. 604-607.
· *Accumulazioni Improprie, Habitat e identità*, Exhibition Catalogue, Customized furnishing of sales spaces, Centro affari e Promozioni Arezzo, July 1992, pp. 64-66.
· Simone Micheli, *Accumulazioni Improprie, Habitat e identità*, in "P.A. Professione Architetto", no. 3, 1992, pp. 60-61.
· Luciana Cuomo and Clara Mantica, *La piazza delle idee,* in "Gap Casa", no. 87, May 1992, pp. 96/115.
Ambizioni e Visioni, edited by Cesare Pergola, Alinea, Florence 1992, pp. 56.
· *Architettura e Spazio Sacro nella Modernità*, Venice Biennale catalogue, Abitare Segesta Cataloghi, Milan, 1992, p. 246.
· *Progetti intorno al corpo*, in "Interni", no. 420, May 1992, pp. 161-166.
· *La dualità della pietra,* in "Interni", no. 421, June 1992, pp. 90-97.
· *La casa delle geometrie*, in "Area", no. 11 September 1992, pp. 68-71.
· Marco Casamonti, *Geometria, regola di Architettura*, in "Area", no. 14, June 1993, p. 6.
· *Dedicato a Roma*, Catalogue, Dromos, Rome, 1993.
· *Metamorfosi delle metamorfosi*, in "Area", no. 18, June 1994, pp. 26-37.
· Matteo Vercelloni, *Nuove Abitazioni in Italia 2, Torre domestica,* edited by Silvio San Pietro, Archivolto, Milan 1995, pp. 22-29.
· Aldo De Poli, *La scena simultanea*, in "Area", no. 22, June 1995, pp. 60-63.
· *Stop Line*, in "WDR", volume 5, 1995, pp. 60-63.
· *Il centro altrove*, Triennale di Milan, Exhibition Catalogue, Electa, 1995 pp. 195 and 216.
· *La riscoperta dei luoghi*, in "l'Arca", no. 96, September 1995, p. 90.
· *Il nuovo Teatro Globale*, edited by Pietro Savorelli, Alinea, Florence 1995.
· *Una Provocatoria precarietà*, in "Interni", no. 458, March 1996, pp. 90-92.
· *Nuova architettura italiana*, at the Heimatmuseum Charlottenburg in Berlin, edited by Ado Franchini, exhibition catalogue, 1996, pp. 14-15.
· *Megacentro de Archea en Lombardia*, in "Arquitectura Viva", no. 47, March-April 1996, pp.9.
· *Centre de loisirs a Bergame, Italie*, in "amc", le moniteur architecture, no. 69, March 1996, pp. 16.
· *Discoteca Stop Line, Curno*, in "Diseño Interior", no. 58, 1996, p. 47
· *L'architetto come sismografo*, Venice Biennale, exhibition catalogue, 1996, pp. 250-253.
· Aldo De Poli, *Studio Archea-Centro per attività ricreative a Curno*, in "d'Architettura", no. 15, March 1996, pp. 18-25.
· *Kult*, in "AIT", no. 458, June 1996, pp. 90-92.
· Carlo Branzaglia, *Discodesign in Italia, Stop Line,* edited by Silvio San Pietro, L'Archivolto, Milan, November 1996, pp. 176-189.

参考文献目录

- *Un'idea per le Murate, di Firenze*, 由Sergio Conti编辑, Electa, 佛罗伦萨, 1989年.
- Luigi Moiraghi, *Due progetti per il Forum di Tokyo*, 在"Arca"之内, 第51期, 1991年7-8月刊, 页号 28-37.
- Gianni Pettena, *In un interno fiorentino*, 在"Domus"项下, 第730期, 1991年9月刊, 页号 4-5.
- *Una porta per Venezia*, 在"Materia"项下, 第8期, 1991年第3季刊, 页号 54-57.
- *L'idea: simmetria e classicità*, 在"Interni"项下, 第416期, 1991年9月刊, 页号 160.
- Antonello Boschi, *Nuovo museo dell'Acropoli di Atene*, Ad Atene piccola!, 在"P.A. Professione Architetto"项下, 第1期, 1992年1月刊, 页号 40-47.
- *Una porta per Venezia*, 在"Phalaris"项下, 第18期, Anno IV, 1992年1-2月刊, 页号 29.
- Marco Casamonti, *Un simbolo antico per nuovo denaro, Recupero di un antico borgo a Todi*, 在"Recuperare"项下, 第7期, 1992年9月刊, 页号 604-607.
- *Accumulazioni Improprie, Habitat e identità*, 展览目录, 销售场所的定制室内陈设, Centro affari e Promozioni Arezzo, 1992年7月刊, 页号 64-66.
- Simone Micheli, *Accumulazioni Improprie, Habitat e identità*, 在"P.A. Professione Architetto"项下, 第3期, 1992年 页号 60-61.
- Luciana Cuomo和Clara Mantica, *La piazza delle idee*, 在"Gap Casa"项下, 第87期, 1992年5月刊, 页号 96/115.

Ambizioni e Visioni, 由Cesare Pergola编辑, Alinea, 佛罗伦萨1992年, 页号 56.

- *Architettura e Spazio Sacro nella Modernità*, 威尼斯双年展目录, Abitare Segesta Cataloghi, 米兰, 1992年, 页号246.
- *Progetti intorno al corpo*, 在"Interni"项下, 第420期, 1992年5月刊, 页号 161-166.
- *Progetti intorno al corpo*, 在"Interni"项下, 第421期, 1992年6月刊, 页号 90-97.
- *Progetti intorno al corpo*, 在"Area"项下, 第421期, 1992年9月刊, 页号 68-71.
- Marco Casamonti, *Geometria, regola di Architettura*, 在"Area"项下, 第14期, 1993年6月刊, 页号6.
- *Dedicato a Roma*, 目录, Dromos, 罗马, 1993年.
- *Metamorfosi delle metamorfosi*, 在"Area"项下, 第18期, 1994年6月刊, 页号 26-37.
- Matteo Vercelloni, *Nuove Abitazioni in Italia 2, Torre domestica*, 由Silvio San Pietro编辑, Archivolto, 米兰 1995年, 页号 22-29.
- Aldo De Poli, *La scena simultanea*, 在"Area"项下, 第22期, 1995年6月刊, 页号 60-63.
- *Stop Line*, 在"WDR"项下, 第5卷, 1995年, 页号 60-63.
- *Il centro altrove*, Triennale di Milan, 展览目录, Electa, 1995年 页号195和216.
- *La riscoperta dei luoghi*, 在"l'Arca"项下, 第96期, 1995年9月, 页号90.
- *Il nuovo Teatro Globale*, 由Pietro Savorelli编辑, Alinea, 佛罗伦萨1995年.
- *Una Provocatoria precarietà*, 在"Interni"项下, 第458期, 1996年3月刊,页号 90-92.
- *Nuova architettura italiana*, 于柏林的Heimatmuseum Charlottenburg博物馆举办, 由Ado Franchini编辑, 展览目录, 1996年, 页号 14-15.
- *Megacentro de Archea en Lombardia*, 在"Arquitectura Viva"项下, 第47期, 1996年3-4月刊, 页号9.
- *Centre de loisirs a Bergame*, Italie, 在"amc"项下, le moniteur architecture, 第69期, 1996年3月刊, 页号 16.
- *Discoteca Stop Line*, Curno, 在"Diseño Interior"项下, 第58期, 1996年, 页号 47
- *L'architetto come sismografo*, 威尼斯双年展, 展览目录, 1996年, 页号 250-253.
- Aldo De Poli, *Studio Archea-Centro per attività ricreative a Curno*, 在"d'Architettura"项下, 第15期, 1996年3

· *Piazza a Merate e Centro Uffici a Firenze*, in "d'Architettura", no. 16, 1996, pp. 10-13.
· *Groot warenhuis wordt dans- en ricreatiecentrum*, in "PI Projekt&Interieur", August 1996, pp. 78-82.
· *Centro ricreativo Stop Line*, in "Diseño Interior", no. 62, 1996, pp. 98-105.
· *Architetti fiorentini della nuova generazione, Archea*, catalogue, Spazio Stella exhibition, Matera, Libria, Melfi, Potenza, 1997.
· *Studio Archea*, exhibition catalogue, publication of projects by studio from 1988 to 1998, Alinea, Florence, 1997.
· *Acciaio corten trasparente Stop Line*, in "l'Arca", no. 113, March 1997, pp. 70-74.
· *Curno Bergamo, Stop Line*, in "l'Arca Plus", no. 14, March 1997, pp. 136-143.
· Maria Giulia Zunino, *Album Italiano*, in "Abitare", no. 367, November 1997, pp. 139-147.
· *Eleven Recent Works, Stop Line*, arco editorial s.a., Barcelona, Spain, 1997, pp. 192-201.
· *Premio nazionale di Architettura Luigi Cosenza 1998*, VI edition, Competition catalogue, award for best project built in 1997-98, p. 38 - p. 82.
· *Studio Archea*, in "Almanacco Casabella 1997-1998, Giovani Architetti Italiani", November 1998, pp. 22-25.
· *Venezia la nuova architettura*, Project for the competition for the new IUAV site in the former refrigerator warehouse area of San Basilio, in the exhibition catalogue published by Skira, 1998.
· *Calenzano, concorso ad inviti per la progettazione del centro direzionale e commerciale del capoluogo*, December 1998, Conti Tipocolor, Calenzano, Florence.
· Mario Caldarelli, *Nuovo centro direzionale e commerciale di Calenzano, Un racconto dolce*, in "l'Arca", no. 132, December 1998, p. 22.
· *Nuova sede dell'IUAV nell'area dei magazzini Frigoriferi a San Basilio, Venezia*, "Casabella" insert, March 1998, p. 24.
· *Studio Archea*, in "Almanacco di Casabella 1998-1999 - Giovani Architetti Italiani", November 1999, pp. 32-35.
· *Misterioso e tecnologico, Stop Line*, in "arcVision", no. 15, November 2000, pp. 56-61
· Domizia Mandolesi, *Nuova Agenzia Spaziale Italiana a Roma*, in "l'Industria delle costruzioni", n. 344, June 2000, p. 62.
· Alessandra De Cesaris, *Casa unifamiliare a Leffe, Bergamo*, in "L'industria delle costruzioni", no. 348, October 2000, pp. 54-62.
· *Maisons de créaterurs - intérieurs italiens 1990-1999, Maison a Costa San Giorgio*, 400 Architectures series, Actes Sud-Federico Motta Editore, Arles, France, June 2001, pp. 22-35.
· Patrizia Mello, *Insediamento artigianale in Versilia*, in "Costruire in Laterizio", no. 83, September-October, 2001, pp. 22-25.
· *Nuove chiese italiane tre-24 progetti per nuove chiese commissionati dalla conferenza episcopale italiana, Parrocchia del Redentore in Modena*, in "Casabella" insert, no. 694, October 2001, pp. 13-15.
· *Studio Archea*, in "Almanacco Casabella 2000-2001, Giovani Architetti Italiani", November 2001, pp. 24-27.
· *Showroom, Galleria Tornabuoni Arte*, edited by Antonello Boschi, Tools series, Federico Motta Editore, Milano, 2001, pp. 20-29.
· *Cafés & Restaurants, Stop Line*, Tools series, teNeuses, pp. 24-29.
· *House in Leffe*, in "Detail", *Building Skins*, 2001, pp. 84-85.
· *Nuevos Conceptos de Vivienda, Casa Leffe*, Loft publications s.l., Barcelona, Spain, 2001, pp. 78-83.
· *Lonely Living, Habitaculos solitarios*, in "Diseño Interior", no. 126, 2002, pp. 50-55.

月刊，页号 18-25.
• *Kult*，在“AIT”项下，第458期，1996年6月刊，页号 90-92.
• Carlo Branzaglia, *Discodesign in Italia, Stop Line*, 由Silvio San Pietro编辑，L'Archivolto, 米兰，1996年11月刊，页号 176-189.
• *Piazza a Merate e Centro Uffici a Firenze*, 在“d'Architettura”项下，第16期，1996年，页号 10-13.
• *Groot warenhuis wordt dans- en ricreatiecentrum*, 在“PI Projekt&Interieur”项下，1996年8月刊，页号 78-82.
• *Centro ricreativo Stop Line*, 在“Diseño Interior”项下，第62期，1996年，页号 98-105.
• *Architetti fiorentini della nuova generazione, Archea*，目录，Spazio Stella展览，Matera, Libria, Melfi, Potenza, 1997年.
• *Studio Archea*, 展览目录，自1988年至1998年间由建筑工作室发表的项目出版物，Alinea, 佛罗伦萨, 1997.
• *Acciaio corten trasparente Stop Line*, 在“l'Arca”项下，第113期，1997年3月刊，页号 70-74.
• *Acciaio corten trasparente Stop Line*, 在“l'Arca Plus”项下，第14期，1997年3月刊，页号 136-143.
• *Maria Giulia Zunino, Album Italiano*, 在“Abitare”项下，第367期，1997年11月刊，页号 139-147.
• *Eleven Recent Works, Stop Line, arco editorial s.a.*, 巴塞罗那, 西班牙, 1997年, 页号 192-201.
• *Premio nazionale di Architettura Luigi Cosenza 1998*, 第VI版, 竞赛目录，最优秀1997-98年度竣工项目奖，第38页–第82页.
• Studio Archea，在*“Almanacco Casabella 1997-1998, Giovani Architetti Italiani”*项下，1998年11月刊, 页号 22-25.
• *Venezia la nuova architettura, San Basilio*前冰库区机关报IUAV建址竞赛项目, 在Skira所出版的展览目录内, 1998.
• *Calenzano, concorso ad inviti per la progettazione del centro direzionale e commerciale del capoluogo*, 1998年12月, Conti Tipocolor, Calenzano, 佛罗伦萨.
• Mario Caldarelli, *Nuovo centro direzionale e commerciale di Calenzano*, Un racconto dolce, in “l'Arca”， 第132期, 1998年12月刊，页号22.
• *Nuova sede dell'IUAV nell'area dei magazzini Frigoriferi a San Basilio*, Venezia, “Casabella”插页, 1998年3月刊, 页号24.
• Studio Archea，在“*Almanacco di Casabella 1998-1999, Giovani Architetti Italiani*”项下，1998年11月刊, 页号 32-35.
• *Misterioso e tecnologico*, Stop Line, 在“arcVision”项下, 第15期, 2000年11月刊, 页号 56-61
• Domizia Mandolesi, *Nuova Agenzia Spaziale Italiana a Roma*, 在“l'Industria delle costruzioni”项下，第344期, 2000年6月刊, 页号62.
• Alessandra De Cesaris, *Casa unifamiliare a Leffe*, Bergamo, 在“L'industria delle costruzioni”项下, 第348期, 2000年10月刊, 页号 54-62.
• *Maisons de créaterurs - intérieurs italiens 1990-1999, Maison a Costa San Giorgio*, 400建筑师系列, Actes Sud-Federico Motta Editore, 阿尔勒, 法国, 2001年6月刊, 页号 22-35.
• *Patrizia Mello, Insediamento artigianale in Versilia*, 在“Costruire in Laterizio”项下, 第83期, 2001年9-10月刊, 页号 22-25.
• *Nuove chiese italiane tre-24 progetti per nuove chiese commissionati dalla conferenza episcopale italiana, Parrocchia del Redentore in Modena*, 在“Casabella”插页, 第694期, 2001年10月刊, 页号 13-15.
• Studio Archea，在“*Almanacco Casabella 2000-2001, Giovani Architetti Italiani*”项下，2001年11月刊, 页号 24-27.
• *Showroom*, Galleria Tornabuoni Arte, 由Antonello

- *Studio Archea, Novoli*, "Casabella" insert, no. 703, March 2002, pp. 52-57.
- *Firenze, area ex Fiat, Novoli*, Exhibition Catalogue, Next 8. International Architecture, Venice Biennial, Edizioni Marsilio, 2002, pp. 160-161.
- *Lonely Living*, Exhibition Catalogue, Next 8. International Architecture, Venice Biennial, Edizioni Marsilio, 2002, pp. 188-189.
- *-40 la Nuova Generazione dell'Architettura Internazionale, Archea - Centro Divertimenti Stop Line*, Skira, May 2002, pp. 102-105.
- *Lonely Living, l'Architettura dello spazio primario - Buonasera Signor Ionda*, Exhibition Catalogue, breakout event at the 8th International Architecture, organized by Aid'A – Agenzia Italiana di Architettura with the Venice Biennial, Federico Motta Editore, Milan September 2002, pp. 74-77.
- *Nuova architettura di Pietra in Italia, Abitazione unifamiliare, Leffe, Bergamo*, 2002, Gruppo editoriale Faenza editrice, Faenza, Ravenna, 2002, pp. 42-47.
- *Dal Futurismo al Futuro possibile nell'Architettura Italiana Contemporanea, Centro Stop Line, Curno, Bergamo 1995*, Exhibition Catalogue, Skira, Geneva-Milan July 2002, pp. 252.
- *Concorso di progettazione per il recupero dell'ex-area Fiat in viale Belfiore a Firenze, Casamonti-Andreini-Turillazzi*, Exhibition Catalogue, Editrice Vallecchi, Florence, October 2002, pp. 58-71.
- *Studio Archea*, in "anc Architecture and Culture", no. 6, 2002, pp. 92-124.
- *Destino di un'opera*, in "Arkitekton", no. 11, Dec.-Jan.-Feb. 2003-2004, pp. 24-26.
- *The Cord*, in "Il Progetto", 2003, pp. 32-35.
- *Abitare poeticamente*, in "Abitare la terra", 2003, pp. 38-39.
- *Progetto per la realizzazione della chiesa del Redentore*, in "Aa", quarterly publication of Order of Architects of Agrigento, February-April 2003, pp. 19-20. *Mad'e, Nuovi Paesaggi, Progetto di via Tirreno a Potenza*, Edizioni progetto nuovo/CC&P with Rubettino Editore, Rome, Spring 2003, p. 9.
- *La nuova Stazione Alta Velocità di Firenze, Gruppo Toscano*, "Casabella" insert, no. 709, March 2003, pp. 72-79.
- Silvia Fabi, *Geologica Showroom*, in "Materia", no. 39, January-April 2003, pp. 44-53.
- *Naturidentitat*, in "AIT", no. 3, March 2003.
- *Pools-Piscine, Archea*, edited by Silvio San Pietro and Paola Gallo, June 2003, L'Archivolto, Milan, pp. 173-175.
- *Enzimi*, in "Area", no. 71, November-December 2003, pp. 172-177.
- Silvia Fabi, *Geologica Stand*, in "Materia", no. 43, January-April 2004, pp. 74-81.
- *Nuova stazione alta velocità*, in "Area", no. 73, March-April 2004, pp. 86-95.
- *La dimensione del Tempo*, in "ar2", no. 3, July-October 2004, pp. 40-49.
- Paolo Di Nardo, *Lo spazio del "tra" l'architettura infrastrutturale nella città contemporanea*, in "AND", no. 2, April 2004, pp. 52-55.
- *Villaggio olimpico a Doha*, in "d'Architettura", no. 24, May-August 2004, pp. 106-113.
- *Nuovo Centro congressi a Trieste*, in "d'Architettura", no. 25, September 2004, pp. 140-143. *Ingresso alla 50ª Mostra d'Arte de La Biennale di Venezia, "the cord"*, with C+S Architetti Associati, in "d'Architettura", no. 25, September 2004, pp. 132-139.
- Silvia Fabi, *Complesso residenziale*, in "Area", no. 76, September-October 2004, pp. 92-101.
- *Schegge di monolite*, in "Casa D", January-February 2005, pp. 76-85.
- *Italy Builds, Discoteca Stop Line*, l'Arca Edizioni,

Boschi编辑, 工具系列, Federico Motta Editore, Milano, 2001, 页号 20-29.

- *Cafés & Restaurants*, Stop Line, 工具系列, teNeuses, 页号 24-29.
- *House in Leffe*, 在“Detail”项下, Building Skins, 2001, 页号 84-85.
- *Nuevos Conceptos de Vivienda*, Casa Leffe, Loft publications s.l., 巴塞罗那, 西班牙, 2001, 页号 78-83.
- *Lonely Living, Habitaculos solitarios*, 在“Diseño Interior”项下, 第126期, 2002, 页号 50-55.
- Studio Archea, *Novoli*, “Casabella”插页, 第703期, 2002年3月刊, 页号 52-57.
- Firenze, area ex Fiat, Novoli, 展览目录, 下一代八大国际建筑（Nxt 8. International Architecture）, 威尼斯双年展, Edizioni Marsilio, 2002, 页号 160-161.
- Lonely Living, 展览目录, 下一代八大国际建筑, 威尼斯双年展, Edizioni Marsilio, 2002, 页号 188-189.
- *-40 la Nuova Generazione dell'Architettura Internazionale*, Archea - Centro Divertimenti Stop Line, Skira, 2002年5月刊, 页号 102-105.
- Lonely Living, *l'Architettura dello spazio primario - Buonasera Signor Ionda*, 展览目录, 第8届国际建筑展突破性事件, 由Aid'A – Agenzia Italiana di Architettura联合威尼斯双年展组织, Federico Motta Editore, 米兰 2002年9月刊, 页号 74-77.
- *Nuova architettura di Pietra in Italia*, Abitazione unifamiliare, Leffe, Bergamo, 2002, Gruppo editoriale Faenza editrice, Faenza, 拉韦纳, 2002, 页号 42-47.
- *Dal Futurismo al Futuro possibile nell'Architettura Italiana Contemporanea*, Centro Stop Line, Curno, Bergamo 1995, 展览目录, Skira, 热那亚–米兰 2002年7月, 页号 252.
- *Concorso di progettazione per il recupero dell'ex-area Fiat in viale Belfiore a Firenze*, Casamonti-Andreini-Turillazzi, 展览目录, Editrice Vallecchi, 佛罗伦萨, 2002年10月, 页号 58-71.
- Studio Archea, 在 “*anc Architecture and Culture*”项下, 第6期, 2002, 页号 92-124.
- *Destino di un'opera*, 在“Arkitekton”项下, 第11期, 12月–1月–2月 2003-2004年, 页号 24-26.
- The Cord, in “Il Progetto”, 2003, 页号 32-35.
- Abitare poeticamente, 在“*Abitare la terra*”项下, 2003, 页号 38-39.
- *Progetto per la realizzazione della chiesa del Redentore*, 在“Aa”项下, 阿格里琴托建筑师协会指令季度出版物, 2003年2月-3月, 页号 19-20.
- *Mad'e, Nuovi Paesaggi, Progetto di via Tirreno a Potenza*, Edizioni progetto nuovo/CC&P with Rubettino Editore, 罗马, 2003春季刊, 页号9.
- *La nuova Stazione Alta Velocità di Firenze*, Gruppo Toscano, “Casabella”插页, 第709期, 2003年3月刊, 页号 72-79.
- Silvia Fabi, *Geologica Showroom*, 在“Materia”项下, 第39期，2003年1–4月刊，页号 44–53.
- *Naturidentitat*, 在“AIT”项下, 第3期，2003年3月刊.
- *Pools-Piscine*, Archea, 由Silvio San Pietro和Paola Gallo编辑, 2003年6月, L'Archivolto, 米兰, 页号 173-175.
- *Enzimi*, 在“Area”项下, 第71期, 2003年11-12月刊, 页号 172-177.
- Silvia Fabi, *Geologica Stand*, 在“Materia”项下, 第43期, 2004年1-4月刊, 页号 74-81.
- *Nuova stazione alta velocità*, 在“Area”项下, 第73期, 2004年3-4月刊, 页号 86-95.
- *La dimensione del Tempo*, 在“ar2”项下, 第3期, 2004年7-10月, 页号 40-49.
- Paolo Di Nardo, *Lo spazio del “tra” l'architettura infrastrutturale nella città contemporanea*, 在“AND”项下, 第2期, 2004年4月刊, 页号 52-55.
- *Villaggio olimpico a Doha*, 在“d'Architettura”项下, 第24期, 2004年5-8月刊, 页号 106-113.
- *Nuovo Centro congressi a Trieste*, 在“d'Architettura”项,

Milan, February 2005, pp. 286-287.
· *Papillon in Venedig*, in "Detail", Bauen mit Stahl, Institut fur internationale Architektur-Documentation, April 2005, pp. 312-315.
· Tommaso Bertini, *Under Wine*, in "AND", no. 4, April 2005, pp. 42-49.
· *Interventi per il nodo di Certosa*, in "Area", no. 79, March-April 2005, pp. 110-119.
· Daria Ricchi, *Antinori Winery, San Casciano*, in "A10", no. 3, May-June 2005, p. 19.
· *2003-2005 Tirana Arkitekture e KonKurse, Archea Studio*, Bashkia e Tiranes, 2005, pp. 110-112.
· Alessandro Napoli, *Studio Archea - intervista a Marco Casamonti*, in "Under 40", otto progetti in parallelo, periodical of the Ordine degli architetti pianificatori, paesaggisti e conservatori della provincia di Salerno, no. 4-5-6 May 2005, pp. 26-37.
· *Biblioteca comunale-Nembro*, in "Area", no. 81, July-August 2005, pp. 78-89.
· *Le Stazioni*, in "Parametro", no. 258-259, July-October 2005, Faenza Editrice, Faenza, Ravenna, pp. 120-121 and pp. 132-133.
· *Verquickt Vielfälting*, in "IdealesHEIM", September 2005, Taschen, pp. 90-93.
· Flores Zanchi, *GranitiFiandre Meeting Room*, in "Materia", no. 48, September-December 2005, pp. 90-96.
· *Archea Associati, Biblioteca Comunale*, in "d'Architettura", no. 28, September-December 2005, pp. 68-69.
· Manolo De Giorgi, *Rimescolare le carte*, in "Speciale Domus" insert, no. 887, December 2005, pp. 10-13.
· Pietro Carlo Pellegrini, *Caffè San Colombano*, in "Materia", no. 49, January-April 2006, pp. 62-69.
· *Milan, Area ex Ansaldo-Bicocca: progetto a quattro mani,* in "Abitare", no. 457, January 2006, p. 32.
· *Architecture and Landscape-Nuova Cantina Antinori*, in "Topos", no. 57, 2006, p. 78.
· *Galleria Tornabuoni Arte*, in "Area", no. 85, March-April 2006, pp. 82-89.
· *Area ex Ansaldo a Milan*, in "AL", Mensile di informazione degli architetti Lombardi, no. 6, June 2006, pp. 12-13.
· *4 Evergreen*, in "Area", no. 86, May-June 2006, Motta Architettura, Milan, pp. 82-89.
· *Archea Associati*, in "Archi 100", no. 41, 2006, pp. 30-33.
· *Archea Associati*, in "Archi 100", no. 44, 2006.
· *Il nuovo palazzo del cinema, concorso internazionale di progettazione, Progetti*, August 2006, Marsilio Editori, Venice, pp. 166-177.
· *Piscine, Piscina in Valseriana*, Tools series, September 2006, Motta Architettura, Milan, pp. 24-31.
· Rita Capezzuto, *Il vino sotto la vigna*, in "Domus", no. 883, December 2006, pp. 62-71.
· *Exhibition Design, Martini, euroluce 2003,* Links, Barcelona, 2006, pp. 38-43.
· Patrizia Valandro, *Nuovo Municipio di Merate*, in "Materia", no. 51, September-December 2006, pp. 134-141.
· *Sardegna i paesaggi del futuro, Studio Archea*, in "Domus" insert, no. 899, January-February 2007, pp. 38-41.
· *Bargino, San Casciano, Nuova Cantina Marchesi Antinori*, in "Europ'A Acciaio Architettura", no. 4, Winter 2007, p. 5.
· *A Touch of New Liberty*, in "A10", no. 13, January-February 2007, p. 50.
· *Museum Für mediterrane nuragische und zeitgenössische*, in "Wettbewerbe aktuell", January 2007, p. 31.
· *New Antinori Winery at Bargino località San Casciano Val di Pesa, Province of Florence*, in "Archicreation", no. 01, China, 2007, pp. 24-33.
· *Caffè e Ristoranti, Caffè San Colombano*, edited by Matteo Genghini e Pasqualino Solomita, Tools series, March 2007, Federico Motta Editore, Milan, pp. 14-21.
· *Loft 2, Studio Archea 1991*, Tools series,

第25期, 2004年9月刊, 页号 140-143.
• *Ingresso alla 50a Mostra d'Arte de La Biennale di Venezia, "the cord"*, 与C+S Architetti Associati, 在"d'Architettura"项下, 第25期, 2004年9月刊, 页号 132-139.
• Silvia Fabi, *Complesso residenziale*, 在"Area"项下, 第76期, 2004年9-10月刊, 页号 92-101.
• *Schegge di monolite*, 在"Casa D"项下, 2005年1-2月刊, 页号 76-85.
• Italy Builds, *Discoteca Stop Line*, l'Arca Edizioni, 米兰, 2005年2月刊, 页号 286-287.
• Papillon in Venedig, 在"Detail"项下, *Bauen mit Stahl, Institut fur internationale Architektur-Documentation*, 2005年4月刊, 页号 312-315.
• Tommaso Bertini, Under Wine, 在"AND"项下, 第4期, 2005年4月刊, 页号 42-49.
• *Interventi per il nodo di Certosa*, 在"Area"项下, 第79号, 2005年3-4月刊, 页号 110-119.
• Daria Ricchi, *Antinori Winery*, San Casciano, 在"A10"项下, 第3期, 2005年5-6月刊, 页号19.
• 2003年-2005年 *Tirana Arkitekture e KonKurse*, Archea Studio, Bashkia e Tiranes, 2005年, 页号 110-112.
• Alessandro Napoli, Studio Archea - intervista a Marco Casamonti, 在"Under 40"项下, *otto progetti in parallelo, periodical of the Ordine degli architetti pianificatori, paesaggisti e conservatori della provincia di Salerno*, 第4-5-6期 2005年5月, 页号 26-37.
• Biblioteca comunale-Nembro, 在"Area"项下, 第81期, 2005年7-8月刊, 页号 78-89.
• *Le Stazioni*, 在"Parametro"项下, 第258-259期, 2005年7-10月刊, Faenza Editrice, 法恩扎, 拉文纳, 页号120-121和页号 132-133.
• *Verquickt Vielfälting*, 在"IdealesHEIM"项下, 2005年9月刊, Taschen, 页号 90-93.
• Flores Zanchi, *GranitiFiandre Meeting Room*, 在"Materia"项下, 第48期, 2005年9-10月刊, 页号 90-96.
• Archea Associati, *Biblioteca Comunale*, 在"d'Architettura"项下, 第28期, 2005年9-12月刊, 页号 68-69.
• Manolo De Giorgi, *Rimescolare le carte*, 在"Speciale Domus"插页内, 第887期, 2005年12月刊, 页号 10-13.
• Pietro Carlo Pellegrini, *Caffè San Colombano*, 在"Materia"项下, 第49期, 2006年1-3月刊, 页号 62-69.
• Milan, *Area ex Ansaldo-Bicocca:progetto a quattro mani*, 在"Abitare"项下, 第457期, 2006年1月刊, 页号32.
• *Architecture and Landscape-Nuova Cantina Antinori*, 在"Topos"项下, 第57期, 2006年, 页号78.
• *Galleria Tornabuoni Arte*, 在"Area"项下, 第85期, 2006年3-4月刊, 页号 82-89.
• *Area ex Ansaldo a Milan*, 在"AL"项下, Mensile di informazione degli architetti Lombardi, 第6期, 2005年6月刊, 页号 12-13.
• *4 Evergreen*, 在"Area"项下, 第86期, 2006年5-6月刊, Motta Architettura, 米兰, 页号 82-89.
• *Archea Associati*, 在"Archi 100"项下, 第41期, 2006年, 页号 30-33.
• *Archea Associati*, 在"Archi 100"项下, 第44期, 2006年.
• *Il nuovo palazzo del cinema, concorso internazionale di progettazione*, Progetti, 2006年8月刊, Marsilio Editori, 威尼斯, 页号 166-177.
• *Piscine*, Piscina in Valseriana, 工具系列, 2006年9月刊, Motta Architettura, 米兰, 页号 24-31.
• Rita Capezzuto, Il vino sotto la vigna, 在"Domus"项下, 第883期, 2006年12月刊, 页号 62-71.
• *Exhibition Design, Martini, euroluce 2003*, Links, 巴塞罗那, 2006年, 页号 38-43.
• Patrizia Valandro, *Nuovo Municipio di Merate*, 在"Materia"项下, 第51期, 2006年9-12月刊, 页号 134-141.
• *Sardegna i paesaggi del futuro*, Studio Archea, 在"Domus"插页内, 第899期, 2007年1-2月刊, 页号 38-41.

Federico Motta Editore, Milan, March 2007, pp. 30-37.
· *Ex teatro Lazzeri e Riqualificazione cinema Metropolitan*, in "Area", no. 91, March-April 2007, pp. 84-95.
· *Ivre de Livres*, in "Archicréé", no. 331, May-June 2007, pp. 78-83.
· *Biblioteca comunale-Nembro*, in "D Casa", March 2008.
· Rita Capezzuto, *La Biblioteca dalle pagine di terracotta*, in "Domus", no. 901, May-June 2007, pp. 78-85.
· Matteo Moscatelli, *Torre delle Arti*, in "Area", no. 92, May-June 2007, pp. 96-107.
· *Discoteche, Stop Line*, Tools series, edited by Silvia Berselli, Federico Motta Editore, Milan, September 2007, pp. 14-21.
· *Città sottili luoghi e progetti di cartone. Lucca 2001-2007, Archea*, Edizioni Libria, Milan, January 2008, pp. 30-31, pp. 86-87, pp. 116-117.
· Alessandra Coppa, *Sunshine 100 Real Estate Cp., LTD - Pechino*, in "Cer Magazine", September 2008, pp. 64-66.
· *Quattro torri per le arti*, in "Geoinforma", no. 1-2008, pp. 21-25.
· Giovanni Polazzi, *Park Albatros Camping*, in "Materia", no. 57, March 2008, pp. 112-121.
Davide Cattaneo, *Libri d'argilla*, in "Archetipo", no. 21, March 2008, pp. 84-95.
· *Topografie, Cantina Antinori*, edited by Fabio Fabbrizzi, April 2008 Alinea Editrice, Florence, pp. 252-257.
· *Manda in cantina un architetto di vino*, in "Panorama", 15 May 2008, p. 233.
· *Biblioteca di Nembro*, in "AL", Mensile di informazione degli architetti Lombardi, no. 6, June 2008, pp. 12-13.
· *L'edificio con le pagine di terracotta*, in "Casaviva", April 2008, Milan.
· *Pagine Mobili*, in "Casamica", no. 3, 2008, pp. 96-62.
· *Progetti per lo sviluppo sostenibile di Shanghai in preparazione di Expo 2010-Better City, Better Life*, in Minambiente-China "Area insert", no. 99, July-August 2008, pp. 28-39.
· *Lagoon Park_Shel[l]ter, Archea Associati and C+S Associati*, Exhibition Catalogue, Side events section, La Biennale 2008 International Architecture Exhibition, Poggibonsi, Siena, Carlo Cambi Editore, 2008.
· Daria Ricchi, *Camping Ground, Rimigliaro Natural Park*, in "A10", no. 22, July-August 2008, p. 24-26.
· *Park Albatros Camping*, in "d'Architettura", no. 36, August 2008, pp. 82-89.
· *Parco Sonoro - Archea Associati*, in "AND", no. 13, September-December 2008, pp. 30-33.
Noemi Cuffia, *En Plein Air*, in "Abitare", no. 487, November 2008, pp. 76-85.
· *Piezas ceramicas rectangulares imitando una estanteria de libros*, in "Detail", Spanish edition, Nov.-Dec. 2008, pp. 1010-1011.
· *Biblioteca Comunale di Nembro*, in "L'industria delle costruzioni", no. 403, November 2008, pp. 46-53.
· *Biblioteca Comunale di Nembro*, in "Solaria", no. 11, December 2008, pp. 48-55.
· *New Library in Nembro, Italy*, in "IW", magazine no. 64, 2008, pp. 96-103.
· *Brit Insurance Designs of the Year, Nembro, Public Library, Nembro, Italy*, exhibition Catalogue, Design Museum, London 2008, p. 16.
· Giovanni Odoni, *Gli interni esterni*, in "Casamica", no. 1-2, 31 January 2009, p. 103.
· *Roma Meno è Più, CMYK- lab Studio Archea*, exhibition Catalogue LISt, gennaio 2009, pp. 94-99.
· *Space Craft2, Nembro Library - San Vincenzo Camping*, Die Gestalten Verlag, Berlin 2009, p. 15 - p. 109.
· *Materials*, in "Contract Project", no, 07, May-June 2009, pp. 40-42.
· *Set in Stone, Screening and filtering the sunlight*, Braun Publishing AG, Berlin 2009, pp. 136-139.
· Laura Andreini, *Archea - Teatro Auditorium del Mggio Musicale Fiorentino,* in "Firenze architettura", May 2009, pp.

· *Bargino, San Casciano, Nuova Cantina Marchesi Antinori,* 在“Europ'A Acciaio Architettura”项下，第4期，2007年冬季刊，页号5.

· *A Touch of New Liberty,* 在“A10”项下，第13期，2007年1-2月刊，页号50.

· *Museum Für mediterrane nuragische und zeitgenössische,* 在“Wettbewerbe aktuell”项下，2007年1月刊，页号31.

· *New Antinori Winery at Bargino località San Casciano Val di Pesa,* 佛罗伦萨省，在“Archicreation”项下，第01期，中国，2007年，页号 24-33.

· *Caffè e Ristoranti, Caffè San Colombano,* 由Matteo Genghini e Pasqualino Solomita编辑，工具系列，2007年3月刊，Federico Motta Editore，米兰，页号 14-21.

· *Loft 2*, Studio Archea 1991，工具系列，Federico Motta Editore，米兰，2007年3月刊，页号 30-37.

· *Ex teatro Lazzeri e Riqualificazione cinema Metropolitan,* 在“Area”项下，第91期，2007年3-4月刊，页号 84-95.

· *Ivre de Livres,* 在“Archicréé”项下，第331期，2007年5-6月刊，页号 78-83.

· *Biblioteca comunale-Nembro,* 在“D Casa”项下，2008年3月刊.

· *Rita Capezzuto*, La Biblioteca dalle pagine di terracotta，在“Domus”项下，第901期，2007年5-6月刊，页号 78-85.

· *Matteo Moscatelli, Torre delle Arti,* 在“Area”项下，第92期，2007年5-6月刊，页号 96-107.

· *Discoteche*, Stop Line，工具系列，由Silvia Berselli编辑，Federico Motta Editore，米兰，2007年9月刊，页号 14-21.

· *Città sottili luoghi e progetti di cartone.Lucca 2001-2007*, Archea, Edizioni Libria，米兰，2008年1月刊，页号30-31，页号86-87，页号 116-117.

· *Alessandra Coppa,* 阳光100房地产公司 - Pechino，在“Cer Magazine”项下，2008年9月刊，页号 64-66.

· *Quattro torri per le arti,* 在“Geoinforma”项下，2008年第1期，页号 21-25.

· Giovanni Polazzi, Park Albatros Camping，在“Materia”项下，第57期，2008年3月刊，页号 112-121.

Davide Cattaneo, *Libri d'argilla,* 在“Archetipo”项下，第21号，2008年3月刊，页号 84-95.

· *Topografie*, Cantina Antinori，由Fabio Fabbrizzi编辑，2008年4月刊 Alinea Editrice，佛罗伦萨，页号 252-257.

· *Manda in cantina un architetto di vino,* 在“Panorama”项下，2008年5月15日，第233页.

· *Biblioteca di Nembro,* 在“AL”项下，*Mensile di informazione degli architetti Lombardi,* 第6期，2008年6月刊，页号 12-13.

· *L'edificio con le pagine di terracotta,* 在“Casaviva”项下，2008年4月刊，米兰.

· *Pagine Mobili,* 在“Casamica”项下，第3期，2008年，页号 96-62.

· *Progetti per lo sviluppo sostenibile di Shanghai in preparazione di Expo 2010*—城市,让生活更美好（Better City, Better Life），在Minambiente–中国“地区插页（Area insert）”，第99期，2008年7-8月刊，页号 28-39.

· *Lagoon Park_Shel[l]ter, Archea Associati and C+S Associati,* 展览目录，侧面事件部分，双年展2008年国际建筑艺术展览，Poggibonsi, Siena, Carlo Cambi Editore, 2008.

· Daria Ricchi, *Camping Ground*, Rimigliaro自然公园，在“A10”项下，第22期，2008年7-8月刊，页号24-26.

· *Park Albatros Camping,* 在“d'Architettura”项下，第36期，2008年8月刊，页号 82-89.

· *Parco Sonoro - Archea Associati,* 在“AND”项下，第13期，2008年9-12月刊，页号 30-33.

Noemi Cuffia, En Plein Air，在“Abitare”项下，第487

42-47.
· *Archea Associati - New Library in Nembro*, in "a+u", no. 465, June 2009, pp. 70-75.
· Giovannl Polazzi, *Archea Associati - Biblioteca e Auditorium Comunali di Curno*, in "Materia", no. 62, June 2009, pp. 86-98.
· Davide Borsa, *Il G8 dalla Sardegna all'Aquila, con la ricostruzione in Abruzzo*, in "Il Giornale dell'Architettura", no. 75, July-August 2009, pp. 86-98.
· *Forme del Moderno*, in "Look lateral", insert in "Marie Claire Maison", July 2009, pp. 25-28.
· *Una struttura Ipogea che valorizza Vino e Paesaggio*, in "Io Architetto", no. 26, July-August 2009, pp. 8-9.
· *Build On, Archea Associati Architects - Ex Teatro Lazzeri-Libreria Edison*, Berlin, Gelstanten, pp. 146-147.
· Guido Incerti, *Progetti contemporanei nel tessuto storico*, in "Architetture Livorno", no. 9, 2009, quarterly architecture magazine, 2009, pp. 16-17.
· *Ex Teatro Lazzeri - Riqualificazione cinema Metropolitan e garage Roma*, in "Architetture Livorno", no. 9, 2009, quarterly architecture magazine, 2009, pp. 44-49.
· *Biblioteca Municipal de Nembro,* in "oficinas", october 2009, pp. 74-79.
· *Gallerie d'Arte, Galleria Tornabuoni Arte - Tornabuoni Portofino Arte*, edited by Chiara Savino, Tools series, 24 ORE Motta Cultura, Milan, September 2009, pp. 300-305 e 306-315.
· *Archea Sustainable Landmarks,* Forma Edizioni, Poggibonsi, Italy, 2009

期, 2008年11月刊, 页号 76-85.
▪ *Piezas ceramicas rectangulares imitando una estanteria de libros*, 在“Detail”项下, 西班牙文版本, 2008年11-12月刊, 页号 1010-1011.
▪ *Biblioteca Comunale di Nembro*, 在“L'industria delle costruzioni”项下, 第403期, 2008年11月刊, 页号 46-53.
▪ *Biblioteca Comunale di Nembro*, 在“Solaria”项下, 第11期, 2008年12月刊, 页号 48-55.
▪ Nembro新图书馆, 意大利, 在“IW”项下, 第64期, 2008, 页号 96-103.
▪ *Brit Insurance*年度设计, Nembro, 公共图书馆, 意大利, 展览目录, Design Museum, 伦敦2008, 页号16.
▪ *Giovanni Odoni, Gli interni esterni*, 在“Casamica”项下, 第1-2期, 2009年1月31日, 页号103.
▪ *Roma Meno è Più, CMYK- lab Studio Archea*, 展览目录清单, gennaio 2009年, 页号 94-99.
▪ *Space Craft2*, Nembro图书馆 – San Vincenzo Camping, Die Gestalten Verlag, Berlin 2009, 页号15 – 页号109.
▪ *Materials*, 在“合同项目(Contract Project)”项下, 第7期, 2009年5-6月刊, 页号 40-42.
▪ 石上扎根, 阳光的遮挡和滤光（Set in Stone, Screening and filtering the sunlight）, Braun Publishing AG, Berlin 2009, 页号 136-139.
▪ Laura Andreini, *Archea - Teatro Auditorium del Mggio Musicale Fiorentino*, 在“Firenze architettura”项下, 2009年5月刊, 页号 42-47.
▪ *Archea Associati – Nembro*的新图书馆, 在“a+u”项下, 第465期, 2009年6月刊, 页号 70-75.
▪ Giovanni Polazzi, *Archea Associati - Biblioteca e Auditorium Comunali di Curno*, 在“Materia”项下, 第62期, 2009年6月刊, 页号 86-98.
▪ Davide Borsa, *Il G8 dalla Sardegna all'Aquila, con la ricostruzione in Abruzzo*, 在“Il Giornale dell'Architettura”项下, 第75期, 2009年7–8月刊, 页号 86-98.
▪ *Forme del Moderno*, 在“横向观察（Look lateral）”项下, “Marie Claire Maison”内插页, 2009年7月刊, 页号 25-28.
▪ *Una struttura Ipogea che valorizza Vino e Paesaggio*, 在“Io Architetto”项下, 第26期, 2009年7–8月刊, 页号 8-9.
▪ Build On, *Archea Associati Architects - Ex Teatro Lazzeri-Libreria Edison*, Berlin, Gelstanten, 页号 146-147.
▪ Guido Incerti, *Progetti contemporanei nel tessuto storico*, 在“Architetture Livorno”项下, 第9期, 2009, 建筑学季刊, 2009年, 页号 16-17.
▪ *Ex Teatro Lazzeri - Riqualificazione cinema Metropolitan e garage Roma*, 在“Architetture Livorno”项下, 第9期, 2009年, 建筑学季刊, 2009年, 页号 44-49.
▪ *Biblioteca Municipal de Nembro*, 在“oficinas”项下, 2009年10月刊, 页号 74-79.
▪ *Gallerie d'Arte, Galleria Tornabuoni Arte - Tornabuoni Portofino Arte*, 由Chiara Savino编辑, 工具系列, 24 ORE Motta Cultura, 米兰, 2009年9月刊, 页号 300-305 e 306-315.
▪ 阿克雅可持续地标, 波吉邦西, 意大利, 2009 年

STAFF 员工

FOUNDING PARTNERS 创建合伙人
Laura Andreini / Marco Casamonti / Giovanni Polazzi

ASSOCIATES 副董事
Silvia Fabi

PARTNERS 合伙人
Marco Gamberi / Francesco Giordani / Li Guojin / Paolo Invidia / Mattia Mugnaini / Michelangelo Perrella / Lara Tonnicchi / Patrizia Valandro

PROJECT LEADERS 项目指导人
Enrico Ancilli / Domenico Giovanni Cacciapaglia / Lorenzo Malavasi / Elena Masci / Alessandro Riccomi / Gabriele Sestini / Wang Xingfang

ARCHITECTS 建筑师
Maria Abbracciavento / Andrea Andreuccetti / Andrea Antonucci / Niccolò Balestri / Mattia Cadenazzi / Luana Carastro / Elena Catalano / Giada Citti / Stefano Costantino / Claudia De Rossi / Enrico De Sanctis / Jin Jin Du / Edward Dumpe / Giovanni Ferrara / Andrea Ferraro / Chiara Francioli / Paolo Gaeta / Marco Giuliotti / Giovanni La Femia / Pablo Lopez Prol / Valentina Malta / Alice Marzorati / Stefano Marcinkiewicz / Federico Mazzoli / Schahrzad Orouji / Marco Orto / Claudio Paba / Federica Poggio / Marco Puppo / Luca Riolfo / Luca Sartori / Massimo Savino / Sara Severini / Yang Shuo / Maria Elena Soliero / Ilaria Stroppa / Tamara Taiocchi / Silvia Tasini / Francesco Volante / Xie Wenxin / Feng Xiancheng / Abduxukur Zayit /

CONSULTING 咨询
Ezio Birondi

GRAPHIC DESIGN 图形设计
Tommaso Bovo / Sara Castelluccio / Vitoria Muzi / Mauro Sampaolesi

EDITORIAL STAFF 编辑人员
Sara Benzi / Maria Giulia Caliri / Katia Carlucci / Caroline Fuchs / Valentina Muscedra / Beatrice Papucci

ADMINISTRATION / SECRETARY 行政管理人员 / 秘书
Liu Bin / Alessandra Laurini / Lucia Platania

PUBLIC RELATIONS 公共关系人员
Antonella Dini

FROM THE BEGINNING 历程的开始

Alessandro Acquaviva / Luigi Andelini / Cecilia Anselmi / Francesco Antonino / Beatrix Arman / Stefano Avesani / Gerilende Back / Laura Badii / Chiara Baracchi / Elisa Baragatti / Francesca Baratto / Andrea Barbierato / Alessandra Barilaro / Emiliano Barneschi / Valentina Baroncini / Erika Bartoli / Ilaria Bartoli / Marcin Batko / Nicola Bellofatto / Fausto Bergamaschi / Alessia Bergamin / Laura Bertolaccini / Cristiano Bianchi / Riccardo Bianchi / Ralf Michael Biele/ Michela Bigagli / Federica Bissocoli / Michelangelo Bonvino / Graziella Bordin / Silvia Borsi / Andrea Boscolo/ Luca Bosetti / Cristina Bradaschia / Davide Brocco / Leonardo Briganti / Daniela Brogi / Ettore Burdese / Emilio Cabella / Elisa Cagelli / Alessia Vitello Margherita Caldi Inchingolo / Anna Calvanese / Alessandro Cambi / Marcella Campa / Sergio Campana / Francesco Capani / Paolo Caratelli / Gregorio Carboni Maestri / Giovanni Carli / Marco Carolei / Silvia Casarotto/ Emilio Casini / Annalisa Castorri / Sofia Cattinari / Mauro Cavallero / Marco Cavalli / Martin Cenek / Maddalena Ceppi / Edoardo Cesaro / Claudio Chiodi / Diana Ciambellotti / Tommaso Cigliana / Mirta Ciari / Gianni Cinali / Giuliano Ciocchetti / Giorgio Cofone / Simone Coletti / Onofrio Colucci / Stefano Conradi / Daniela Corradino / Maria Angela Corsi / Giuliano Cosi / Carlotta Costantino / Alexander Cottaris / Anita Cova / Stefania Crivelli / Licia D'Anella / Donatello D'Angelo / Maja Dapcevic / Cosimo Damiano D'Aprile / Sara Daddario / Chiara Dal Piaz / Andrea Debaggis / Antonella De Bonis / Federica Del Fabbro / Laura Della Badia / Michele Della Vecchia / Marco Del Puglia / Pedro Marques De Sousa / Ella Del Monaco / Lucilla Del Santo / Andrea Destro / Davide Di Franco / Mirko Di Lanzo / Anna Di Napoli / Nunziastella Dileo / Nunzia Di Molfetta / Elisa Di Rosa / Manuel Di Stefano / Domenico Dimichino / Federica Doglio / Mauro Donadello / Stefano D'Ottavio/ Katrin Chantal Dummler / Alessadro Evangelisti / Lorenzo Evangelisti / Maurizio Fagiuoli / Antonella Fantozzi / Francesco Farnetani / Lisa Fellini / Davide Ferrando / Raffaele Ferrandino / Francesca Ferrarini / Giuseppe Fioroni / Alessio Forlano / Marita Formaggio / Giulio Forte / Daniele Franceschin / Paolo Frongia / Francesca Galasso / Davide Gamba / Federica Gargani / Mario Gaudio / Fabio Gammacurta / Jessica Gastaldo / Simone Gavazzeni / Alessandro Gazzoni / Claudia Gelosa / Alessandro Gervasi / Aldo Giacchetto / Jacopo Maria Giagnoni / Alessandro Jovine / Tang Junze / Massimiliano Giberti / Nicola Giuiliani / Simone Gobbo / Pablo Gomez / Lorenzo Gondoni / Bettina Gori / Stefano Grande / Eleonora Grassi / Antonio Greco / Michael Heffernan / Sandra Horesch / Guido Incerti / Barbara Incorvaia / Rebecca Innocenti / Serena Jaff / Yu Ju Zi / Lui Kai / Simone Lapenta / Mauro Lazzari / Raffaella Lecchi / Vincenzo Lechiancole / Duilio Leonio / Sergio Lemma / Nicola Licari / Daria Longinotti / Giuseppe Lorusso / Bia Lucido / Marco Lupi / Raffaella Macci / Jim Mc Charty / Antonella Magherini / Fabiane Maiochi / Massimo Malatesta / Marcello Marchesini / Daniele Marcotulli / Andrea Marelli / Giulia Marino / Loredana Marino / Valentina Martino/ Silvia Mastrovito / David Mata / Jessica Menini / Caterina Micucci / Riccardo Miselli / Alessandra Moncini / Lucrezia Montalbò / Philippe Morel / Lucia Moresi / Isabella Mori / Matteo Moscatelli / Francesca Mura / Eugenia Murialdo / Hu Na / Ren Na / Michael Nardi Tonet / Matteo Negrin / Giuseppina Nerli / Monica Niccolini/ Lidia Nunzi / Antonio Ottomanelli / Patrizia Padula / Mauro Pagnani / Francesca Pagnoncelli / Pierluigi Pala / Saverio Panata/ Annarita Papeschi / Laura Parenti / Gianna Parisse / Sebastiana Patania / Maria Roberta Pansini / Fiorella Pascarella / Antonio Paternostro / Samantha Patroncini / Rosario Patti / Matthew Peek / Francesca Pellegrini / Giulia Pellegrini / Davide Penserini / Lucia Petrelli / Domenico Petruzzi / Giorgia Pezzolla/ Martina Piccolo / Priscilla Pieralli / Serena Pietrantoni / Aleksandar Petrov / Veronica Pirazzini /Alice Poggesi / Valentina Poggi / Massimo Poli / Carlo Prati / Gabriele Pinca / Alessia Pincini / Francesca Privitera / Julien Proietti Peparelli / Francesco Pergetti / Massimo Pironnello / Marigabriela Puerta / Laura Puliti / Beatrice Quagliotti / Antonella Radicchi / Francesco Renieri / Matteo Raselli / Reana Reale / Lucia Repetto / Karin Revoltella / Daria Ricchi / Rocco Ricciardi / Diego Rizza / Antonello Rizzo / Marika Roccabruna / Ivano Rocco / Luca Romagnoli / Emiliano Romanazzi / Mario Romano / David Moreno Romero / Roberto Rontini / Andrea Rossi / Pierpaolo Rossi / Camilla Rossi / Chiara Sabattini / Tommaso Sangaino / Nicola Santini / Francesca Savi / Maarten Scheurwater / Giampietro Sciarpa / Roberto Secchi / Annalisa Selvi /Antonella Serra / Davide Servente / Fabio Sgaramella / Qui Shuangci / Antonio Silvestri / Angela Simonelli / Cristina Sipolo / Ruggero Sozzi / Simone Speciale / Giuseppe Pezzano / Rosy Strati / Pier Paolo Taddei / Fulvia Tallini / Chiara Tambani / Michelangelo Tiefenthaler / Lian Ting / Riccardo Todesco / Francesco Tomatis / Nicoletta Tona / Cristiano Torre / Fiamma Tortoli / Luisa Troncanetti / Stefano Tronci / Antonella Tundo / Chiara Urciuolo / Valentina Valentini / Vito Viesti / Marco Vimercati / Giusi Viola / Marco Visicaro / Wang Wei / Wei Ye / Feng Yi / Flores Zanchi / Claudio Zappia / Francesca Zeri / Silvia Zini / Lorenzo Zoli / Francesca Zorzetto / Marco Zuttioni

O M J A

A Z M S X

A R C H E A

C E T R X

M V I A R D

致谢

A&B Photodesign 198-199, 204-205, 206, 208-209, 210, 211

Dario Bertuzzi 534-535

Tommaso Bovo 646-647, 650

Alessandra Chemollo 536-537, 538-539, 540-541

Alessandro Ciampi 142, 146-147, 360-363, 378-379, 396-397, 399, 400-401, 402-403, 404, 405, 406, 407, 408, 409, 428-429, 430, 434, 463, 516-517, 768-769, 770, 772, 773, 780-781, 810-811, 813-814, 815, 816-817, 818, 820

Saverio Lombardi Vallauri, Pietro Savorelli 354-355, 357, 358-359, 360-361, 362, 363, 364-365, 366, 367

Moreno Maggi 371, 372-373, 374-375

Valentina Muscedra 102-103, 108-109, 110-111, 112, 114-115, 645

Alessandro Riccomi 514-515, 518, 621, 522-523

Christian Richters 508-509, 510

Pietro Savorelli 76-77, 80-81, 82-83, 89, 90-91, 92, 93, 94-95, 102-103, 106-107, 108-109, 116, 118-119, 122, 123, 124-125, 212-213, 215, 260-261, 262-263, 266-267, 270-271, 272-273, 274-275, 276-277, 278-279, 282, 283, 284, 286, 287, 288, 289, 290-291, 294-295, 376-377, 380-381, 383, 384-385, 386, 387, 388-389, 390-391, 393, 394-395, 396-397, 399, 400, 402, 403, 426, 427, 440-441, 446, 449, 450, 452, 453, 454-455, 456, 457, 468-469, 485, 472-473, 474-475, 490-491, 492-493, 495, 496, 499, 500, 502-503, 504, 506, 507, 511, 512-513, 517, 518-519, 521, 522-523, 524-525, 527, 528-529, 530-531, 560-561, 563, 564-565, 566-567, 568, 569, 571, 572-573, 574, 575, 576, 579, 580, 644, 750-751, 753, 754, 755, 774-775, 777, 778, 779, 786-787, 788, 789, 792-793, 794, 795, 796-797, 799, 801, 804-805, 807, 809, 821, 822-823, 824-825, 826, 827, 828-829, 831-832, 833, 835, 836-837, 838-839, 841-842, 843, 844-845, 846, 847, 849, 850-851

Antonio Ottomanelli 648-649, 906

Charlie Xia 650-651, 654-655, 657, 658-659, 660-661, 662, 663, 664, 666-667, 668-669

Archivio Archea 79, 84-85, 202-203, 216-217, 218-219, 546, 580-581, 622-623, 625, 626-627, 628-629, 631, 632, 633, 645

THANKS

There are a great number of people to thank, as is easy to imagine, for their contributions over our more than twenty years history of designing objects and homes. Listing them one by one, we run the risk of leaving out some people, even those who were very significant. Yet, this is a risk we will have to take in the hope that those who know they have helped us, in so many different ways, in the enormous task of trying to leave little pieces of art and architecture, useful for dwelling and thinking about its real meaning, will understand that our possible oversight takes nothing away from their invaluable contribution.
Our pursuits, never seen as mere work, are fueled by every piece of input, every small contribution, every building material and human effort. As we have moved between our university desks and our design desks, in a ceaseless back and forth, from the beginning and to this day, we have found it amazing that to do that which we studied for, that which is our passion and our raison d'être, there are people willing to pay us, corresponding a value to our work in money, which has sincerely never interested us.
On a cultural level, there are many important individuals who have fundamentally shaped our education, who have been mentioned and described in the "dialogue" in the previous pages; however, as that was part of a recent interview only with Marco Casamonti, we should note that Giovanni Polazzi was given his architectural and human education under the aegis of Pier Luigi Spadolini, who is unquestionably owed profound thanks, though he is no longer with us other than through his questions and his teachings.
To our clients, whom we have never considered as such, but as friends and co-creators of the design; from those who were most difficult to those who were the most interested and understanding, all deserve major recognition, as nothing in these pages would have been possible without them. Their names and backgrounds are listed in the details that go with each project, though it was unfortunately impossible to publish all our projects.
Much thanks to Adriano, Alban, Albiera, Alda, Alessandro, Andrea, Angelo, Claudio, Donato, Elena, Enrico, Franco, Guido, Mario, Massimo, Piero, Roberto, Valerio, Vanni.
To every architect and individual who has collaborated with us throughout the years, whose names are listed in "from the beginning", we ought to erect a monument dedicated to the hard work and commitment that they have given us, supporting us even in our most intense, difficult times.
Thank you to the studios and architects who have worked with us on many, many projects.
And to those who made it possible, in practical terms, to put out this book, which seemed as if it would never get out the door. Thanks to Daniela Brogi for having written all the text, to Fulvio Gallotti, who corrected the images and for his incomparable photo editing, and Carlo Cambi who printed them so quickly and brought an end to our constant rethinkings.
Once again, thank you everyone.

鸣谢

这里要向许多人致谢，我们很容易想像他们在20多年以来在设计各类对象和住所方面所做出的贡献。即使我们逐一列出各位，也会冒着遗漏掉一些人的风险，即使他们的贡献非常巨大。但我们仍然要冒这个险，希望那些知道他们帮助过我们的人士理解，我们可能的遗漏完全不会减低他们所做出的宝贵贡献的价值，在那些为了尝试留下少数能够让我们居住并思考人生真正意义的艺术和建筑而执行的数量巨大的任务当中，他们曾经以许多不同的方式帮助我们。我们的追求绝对不能只视作工作而已，这个追求在每一个信息投入、每一个小小贡献、每一种建筑材料和大家所付出心血的推动下前行。在我们穿梭于大学教席与设计工作台之间，在从不间断的反复思考之间，从开始创建到今天，我们惊奇地发现，我们所做的工作正是我们所学习的学科，而这些正是我们的热情所在以及我们存在的理由，而人们愿意向我们支付酬劳，用金钱来给我们的工作标上价值的事实，却从来没有真正地让我们感过兴趣。从文化的层面来说，有许多重要的人物都从根本上塑造了我们的教育，前些页的“对话”部分已经提到并描述了他们；但是，由于这部分来自于最近对只针对Marco Casamonti的访谈，我们应当注意到，Giovanni Polazzi将自己的建筑和为人的教育归功于Pier Luigi Spadolini，并对他表示了无庸置疑的深刻感谢，虽然除了通过他的提问和他的教学之外，他已经不在我们身边了。对我们的客户，我们从来没有把他们当作客户，而是当作朋友和设计方案的共同创造者；从那些最严格要求的客户到那些最有趣和最能理解的客户，都值得我们大力感谢，如果没有了他们，这里所谈到的一切都不会存在。他们的名字和背景都在每个项目当中详细列出，尽管我们很不幸地不可能将所有项目都发表出来。我们深深地感谢Adriano、Alban、Albiera、Alda、Alessandro、Andrea、Angelo、Claudio、Donato、Elena、Enrico、Franco、Guido、Mario、Massimo、Piero、Roberto、Valerio、Vanni。对这些年来曾与我们合作的每一位建筑师和个人，他们的名字列出“历程的开始”部分，对于他们向我们付出的辛勤劳动和投入，为他们即使在我们最紧张、最困难时刻对我们的支持，我们应当而为他们建立一座纪念碑。对一直以来在许多、许多项目上与我们合作的建筑师事务所和建筑师，我们深致谢意。我们感谢那些实实在在地让本书得以付印的人们，感谢他们长时间的埋头苦干。感谢Daniela Brogi编写了全部正文，感谢Fulvio Gallotti修正了图像和无比伦比的照片编辑工作，感谢Carlo Cambi如此之愉地将其付印并终结了我们总停不下来的反复思量。再一次地，感谢你们所有人。

图书在版编目（CIP）数据

可持续性地标建筑 : 汉英对照 / 石大伟 主编. --北京 : 中国林业出版社, 2012.6

ISBN 978-7-5038-6603-6

Ⅰ. ①可… Ⅱ. ①石… Ⅲ. ①建筑设计－作品集－世界－现代 Ⅳ. ①TU206

中国版本图书馆CIP数据核字(2012)第094637号

可持续性地标建筑（下） 石大伟 主编

责任编辑：李 顺
出版咨询：（010）83223051

出 版：中国林业出版社（100009 北京西城区德内大街刘海胡同7号）
印 刷：北京时捷印刷有限公司
发 行：新华书店北京发行所
电 话：（010）83224477
版 次：2012年6月第1版
印 次：2012年6月第1次
开 本：787mm×1092mm 1／12
印 张：76
字 数：200千字
定 价：880.00元（上、中、下册）